THE FUTURE OF COMPUTING PERFORMANCE

Game Over or Next Level?

Samuel H. Fuller and Lynette I. Millett, *Editors*

Committee on Sustaining Growth in Computing Performance

Computer Science and Telecommunications Board

Division on Engineering and Physical Science

NATIONAL RESEARCH COUNCIL
OF THE NATIONAL ACADEMIES

THE NATIONAL ACADEMIES PRESS
Washington, D.C.
www.nap.edu

THE NATIONAL ACADEMIES PRESS   500 Fifth Street, N.W.   Washington, DC 20001

NOTICE: The project that is the subject of this report was approved by the Governing Board of the National Research Council, whose members are drawn from the councils of the National Academy of Sciences, the National Academy of Engineering, and the Institute of Medicine. The members of the committee responsible for the report were chosen for their special competences and with regard for appropriate balance.

Support for this project was provided by the National Science Foundation under award CNS-0630358. Any opinions, findings, conclusions, or recommendations expressed in this publication are those of the authors and do not necessarily reflect the views of the organization that provided support for the project.

International Standard Book Number-13:   978-0-309-15951-7
International Standard Book Number-10:   0-309-15951-2
Library of Congress Control Number: 2011923200

Additional copies of this report are available from

The National Academies Press
500 Fifth Street, N.W., Lockbox 285
Washington, D.C. 20055
800 624-6242
202 334-3313 (in the Washington metropolitan area)
http://www.nap.edu

Copyright 2011 by the National Academy of Sciences. All rights reserved.

Printed in the United States of America

THE NATIONAL ACADEMIES
Advisers to the Nation on Science, Engineering, and Medicine

The **National Academy of Sciences** is a private, nonprofit, self-perpetuating society of distinguished scholars engaged in scientific and engineering research, dedicated to the furtherance of science and technology and to their use for the general welfare. Upon the authority of the charter granted to it by the Congress in 1863, the Academy has a mandate that requires it to advise the federal government on scientific and technical matters. Dr. Ralph J. Cicerone is president of the National Academy of Sciences.

The **National Academy of Engineering** was established in 1964, under the charter of the National Academy of Sciences, as a parallel organization of outstanding engineers. It is autonomous in its administration and in the selection of its members, sharing with the National Academy of Sciences the responsibility for advising the federal government. The National Academy of Engineering also sponsors engineering programs aimed at meeting national needs, encourages education and research, and recognizes the superior achievements of engineers. Dr. Charles M. Vest is president of the National Academy of Engineering.

The **Institute of Medicine** was established in 1970 by the National Academy of Sciences to secure the services of eminent members of appropriate professions in the examination of policy matters pertaining to the health of the public. The Institute acts under the responsibility given to the National Academy of Sciences by its congressional charter to be an adviser to the federal government and, upon its own initiative, to identify issues of medical care, research, and education. Dr. Harvey V. Fineberg is president of the Institute of Medicine.

The **National Research Council** was organized by the National Academy of Sciences in 1916 to associate the broad community of science and technology with the Academy's purposes of furthering knowledge and advising the federal government. Functioning in accordance with general policies determined by the Academy, the Council has become the principal operating agency of both the National Academy of Sciences and the National Academy of Engineering in providing services to the government, the public, and the scientific and engineering communities. The Council is administered jointly by both Academies and the Institute of Medicine. Dr. Ralph J. Cicerone and Dr. Charles M. Vest are chair and vice chair, respectively, of the National Research Council.

www.national-academies.org

COMMITTEE ON SUSTAINING GROWTH IN COMPUTING PERFORMANCE

SAMUEL H. FULLER, Analog Devices Inc., *Chair*
LUIZ ANDRÉ BARROSO, Google, Inc.
ROBERT P. COLWELL, Independent Consultant
WILLIAM J. DALLY, NVIDIA Corporation and Stanford University
DAN DOBBERPUHL, P.A. Semi
PRADEEP DUBEY, Intel Corporation
MARK D. HILL, University of Wisconsin–Madison
MARK HOROWITZ, Stanford University
DAVID KIRK, NVIDIA Corporation
MONICA LAM, Stanford University
KATHRYN S. McKINLEY, University of Texas at Austin
CHARLES MOORE, Advanced Micro Devices
KATHERINE YELICK, University of California, Berkeley

Staff

LYNETTE I. MILLETT, Study Director
SHENAE BRADLEY, Senior Program Assistant

COMPUTER SCIENCE AND TELECOMMUNICATIONS BOARD

ROBERT F. SPROULL, Sun Labs, *Chair*
PRITHVIRAJ BANERJEE, Hewlett Packard Company
STEVEN M. BELLOVIN, Columbia University
WILLIAM J. DALLY, NVIDIA Corporation and Stanford University
SEYMOUR E. GOODMAN, Georgia Institute of Technology
JOHN E. KELLY, III, IBM
JON M. KLEINBERG, Cornell University
ROBERT KRAUT, Carnegie Mellon University
SUSAN LANDAU, Radcliffe Institute for Advanced Study
PETER LEE, Microsoft Corporation
DAVID LIDDLE, US Venture Partners
WILLIAM H. PRESS, University of Texas
PRABHAKAR RAGHAVAN, Yahoo! Research
DAVID E. SHAW, Columbia University
ALFRED Z. SPECTOR, Google, Inc.
JOHN SWAINSON, Silver Lake Partners
PETER SZOLOVITS, Massachusetts Institute of Technology
PETER J. WEINBERGER, Google, Inc.
ERNEST J. WILSON, University of Southern California

JON EISENBERG, Director
RENEE HAWKINS, Financial and Administrative Manager
HERBERT S. LIN, Chief Scientist
LYNETTE I. MILLETT, Senior Program Officer
EMILY ANN MEYER, Program Officer
ENITA A. WILLIAMS, Associate Program Officer
VIRGINIA BACON TALATI, Associate Program Officer
SHENAE BRADLEY, Senior Program Assistant
ERIC WHITAKER, Senior Program Assistant

For more information on CSTB, see its website at
http://www.cstb.org, write to CSTB, National Research Council,
500 Fifth Street, N.W., Washington, D.C. 20001, call (202) 334-2605,
or e-mail the CSTB at cstb@nas.edu.

Preface

Fast, inexpensive computers are now essential for nearly all human endeavors and have been a critical factor in increasing economic productivity, enabling new defense systems, and advancing the frontiers of science. But less well understood is the need for *ever-faster* computers at ever-lower costs. For the last half-century, computers have been doubling in performance and capacity every couple of years. This remarkable, continuous, exponential growth in computing performance has resulted in an increase by a factor of over 100 per decade and more than a million in the last 40 years. For example, the raw performance of a 1970s supercomputer is now available in a typical modern cell phone. That uninterrupted exponential growth in computing throughout the lifetimes of most people has resulted in the expectation that such phenomenal progress, often called Moore's law, will continue well into the future. Indeed, societal expectations for increased technology performance continue apace and show no signs of slowing, a trend that underscores the need to find ways to sustain exponentially increasing performance in multiple dimensions.

The essential engine that made that exponential growth possible is now in considerable danger. Thermal-power challenges and increasingly expensive energy demands pose threats to the historical rate of increase in processor performance. The implications of a dramatic slowdown in how quickly computer performance is increasing—for our economy, our military, our research institutions, and our way of life—are substantial. That obstacle to continuing growth in computing performance is by now well understood by the designers of microprocessors. Their initial response was to design multiprocessor (often referred to as multicore) chips, but fundamental challenges in algorithm and software design limit the widespread use of multicore systems.

Even as multicore hardware systems are tailored to support software that can exploit multiple computation units, thermal constraints will continue to be a primary concern. It is estimated that data centers delivering Internet services consume over 1.5 percent of U.S. electric power. As the use of the Internet continues to grow and massive computing facilities are demanding that performance keep doubling, devoting corresponding increases in the nation's electrical energy capacity to computing may become too expensive.

We do not have new software approaches that can exploit the innovative architectures, and so sustaining performance growth—and its attendant benefits—presents a major challenge. The present study emerged from discussions among members of the Computer Science and Telecommunications Board and was sponsored by the National Science Foundation. The original statement of task for the Committee on Sustaining Growth in Computing Performance is as follows:

> This study will bring together academic and industry researchers, application developers, and members of the user community to explore emerging challenges to sustaining performance growth and meeting expectations in computing across the broad spectrum of software, hardware, and architecture. It will identify key problems along with promising emerging technologies and models and describe how these might fit together over time to enable continued performance scaling. In addition, it will focus attention on areas where there are tractable problems whose solution would have significant payback and at the same time highlight known solutions to challenges that already have them. The study will outline a research, development, and educational agenda for meeting the emerging computing needs of the 21st century.

Parallelism and related approaches in software will increase in importance as a path to achieving continued performance growth. There have been promising developments in the use of parallel processing in some scientific applications, Internet search and retrieval, and the processing of visual and graphic images. This report reviews that progress and recommends subjects for further research and development. Chapter 1 examines the need for high-performance computers, and computers that are increasingly higher-performing, in a variety of sectors of society. The need may be intuitively obvious to some readers but is included here to be explicit about the need for continued performance growth. Chapter 2 examines the aspects of "performance" in depth. Often used as shorthand for speed, performance is actually a much more multidimensional concept. (Appendix A provides a brief history of computing performance as a complement to Chapter 2.) Chapter 3 delves into the fundamental reasons why single-processor performance has stopped its dramatic, exponential growth and why this is a fundamental change rather than a temporary nuisance. Chapter 4 addresses the fundamental challenge now facing the computer science and engineering community: how to exploit parallelism in software and hardware. Chapter 5 outlines the committee's recommended research, practice, and education agenda to meet those challenges.

This report represents the cooperative effort of many people. The members of the study committee, after substantial discussions, drafted and worked though several revisions of the report. We particularly appreciate the insights and perspectives provided by the following experts who briefed the committee:

Jeff Dean, Google,
Robert Doering, Texas Instruments,
Michael Foster, National Science Foundation,
Garth Gibson, Carnegie Mellon University,
Wen-Mei Hwu, University of Illinois at Urbana-Champaign,
Bruce Jacob, University of Maryland,
Jim Larus, Microsoft,
Charles Leiserson, Massachusetts Institute of Technology,
Trevor Mudge, University of Michigan,
Daniel Reed, Microsoft,
Phillip Rosedale, Linden Lab,
Vivek Sarkar, Rice University,
Kevin Skadron, University of Virginia,
Tim Sweeny, Epic Games, and
Tom Williams, Synopsys.

The committee also thanks the reviewers who provided many perceptive comments that helped to improve the content of the report materially. The committee thanks Michael Marty, who worked with committee member Mark Hill to update some of the graphs, and Paul S. Diette of the Diette Group, who assisted in refining the images. The committee appreciates the financial support provided by the National Science Foundation. The committee also gratefully acknowledges the assistance of members of the National Research Council staff. Lynette Millett, our study director, ably served the critical roles of study organizer, report editor, and review coordinator. Jon Eisenberg provided many valuable suggestions that improved the quality of the final report.

It is difficult to overstate the importance of ever-more-capable computers to the U.S. industrial and social infrastructure, economy, and national security. The United States cannot afford to let this growth engine stall out, and a concerted effort is needed to sustain it. Several centers for parallel computing have already been established in leading research universities. Those centers are a good start, and additional, strong actions are required in many subdisciplines of computer science and computer engineering. Our major goal for this study is to help to identify the actions and opportunities that will prove most fruitful.

> Samuel H. Fuller, *Chair*
> Committee on Sustaining Growth
> in Computing Performance

Acknowledgment of Reviewers

This report has been reviewed in draft form by individuals chosen for their diverse perspectives and technical expertise, in accordance with procedures approved by the National Research Council's (NRC's) Report Review Committee. The purpose of this independent review is to provide candid and critical comments that will assist the institution in making its published report as sound as possible and to ensure that the report meets institutional standards for objectivity, evidence, and responsiveness to the study charge. The review comments and draft manuscript remain confidential to protect the integrity of the deliberative process. We wish to thank the following individuals for their review of this report:

Tilak Agerwala, IBM Research,
David Ceperley, University of Illinois,
Robert Dennard, IBM Research,
Robert Doering, Texas Instruments, Inc.,
Urs Hölzle, Google, Inc.,
Norm Jouppi, Hewlett-Packard Laboratories,
Kevin Kahn, Intel Corporation,
James Kajiya, Microsoft Corporation
Randy Katz, University of California, Berkeley,
Barbara Liskov, Massachusetts Institute of Technology,
Keshav Pingali, University of Texas, Austin,
James Plummer, Stanford University, and
Vivek Sarkar, Rice University.

Although the reviewers listed above have provided many constructive comments and suggestions, they were not asked to endorse the conclusions or recommendations, nor did they see the final draft of the report before its release. The review of this report was overseen by Butler Lampson, Microsoft Corporation. Appointed by the National Research Council, he was responsible for making certain that an independent examination of this report was carried out in accordance with institutional procedures and that all review comments were carefully considered. Responsibility for the final content of this report rests entirely with the authoring committee and the institution.

Contents

ABSTRACT   1

SUMMARY   5

1   THE NEED FOR CONTINUED PERFORMANCE GROWTH   21
Why Faster Computers Are Important, 22
The Importance of Computing Performance for the Sciences, 29
The Importance of Computing Performance for Defense and National Security, 36
The Importance of Computing Performance for Consumer Needs and Applications, 44
The Importance of Computing Performance for Enterprise Productivity, 47

2   WHAT IS COMPUTER PERFORMANCE?   53
Why Performance Matters, 58
Performance as Measured by Raw Computation, 59
Computation and Communication's Effects on Performance, 62
Technology Advances and the History of Computer Performance, 65
Assessing Performance with Benchmarks, 68
The Interplay of Software and Performance, 70
The Economics of Computer Performance, 75

3 POWER IS NOW LIMITING GROWTH IN COMPUTING   80
   PERFORMANCE
   Basic Technology Scaling, 83
   Classic CMOS Scaling, 84
   How CMOS-Processor Performance Improved Exponentially,
      and Then Slowed, 87
   How Chip Multiprocessors Allow Some Continued
      Performance-Scaling, 90
   Problems in Scaling Nanometer Devices, 94
   Advanced Technology Options, 97
   Application-Specific Integrated Circuits, 100
   Bibliography, 103

4 THE END OF PROGRAMMING AS WE KNOW IT   105
   Moore's Bounty: Software Abstraction, 106
   Software Implications of Parallelism, 110
   The Challenges of Parallelism, 116
   The State of the Art of Parallel Programming, 119
   Parallel-Programming Systems and the Parallel
      Software "Stack," 127
   Meeting the Challenges of Parallelism, 130

5 RESEARCH, PRACTICE, AND EDUCATION TO MEET   132
   TOMORROW'S PERFORMANCE NEEDS
   Systems Research and Practice, 133
   Parallel-Programming Models and Education, 146
   Game Over or Next Level? 150

APPENDIXES

A A History of Computer Performance   155
B Biographies of Committee Members and Staff   160
C Reprint of Gordon E. Moore's "Cramming More Components
   onto Integrated Circuits"   169
D Reprint of Robert H. Dennard's "Design of Ion-Implanted
   MOSFET's with Very Small Physical Dimensions"   174

Abstract

Information technology (IT) has the potential to continue to dramatically transform how we work and live. One might expect that future IT advances will occur as a natural continuation of the stunning advances that IT has enabled over the last half-century, but reality is more sobering.

IT advances of the last half-century have depended critically on the rapid growth of single-processor performance—by a factor of 10,000 in just the last 2 decades—at ever-decreasing cost and with manageable increases in power consumption. That growth stemmed from increasing the number and speed of transistors on a processor chip by reducing their size and—with improvements in memory, storage, and networking capacities—resulted in ever more capable computer systems. It was important for widespread IT adoption that the phenomenal growth in performance was achieved while maintaining the *sequential stored-program model* that was developed for computers in the 1940s. Moreover, computer manufacturers worked to ensure that specific instruction set compatibility was maintained over generations of computer hardware—that is, a new computer could run new applications, and the existing applications would run faster. Thus, software did not have to be rewritten for each hardware generation, and so ambition and imagination were free to drive the creation of increasingly innovative, capable, and computationally intensive software, and this in turn inspired businesses, government, and the average consumer to buy successive generations of computer software and hardware. Software and hardware advances fed each other, creating a virtuous IT economic cycle.

Early in the 21st century, improvements in single-processor performance slowed, as measured in instructions executed per second, and such performance now improves at a very modest pace, if at all. This abrupt shift is due to fundamental limits in the power efficiency of complementary metal oxide semiconductor integrated circuits (used in virtually all computer chips today) and apparent limits in the efficiencies that can be exploited in single-processor architectures. Reductions in transistor size continue apace, and so more transistors can still be packed onto chips, albeit without the speedups seen in the past. As a result, the computer-hardware industry has commenced building chips with multiple processors. Current chips range from several complex processors to hundreds of simpler processors, and future generations will keep adding more. Unfortunately, that change in hardware requires a concomitant change in the software programming model. To use chip multiprocessors, applications must use a *parallel programming model*, which divides a program into parts that are then executed in parallel on distinct processors. However, much software today is written according to a sequential programming model, and applications written this way cannot easily be sped up by using parallel processors.

The only foreseeable way to continue advancing performance is to match parallel hardware with parallel software and ensure that the new software is portable across generations of parallel hardware. There has been genuine progress on the software front in specific fields, such as some scientific applications and commercial searching and transactional applications. Heroic programmers can exploit vast amounts of parallelism, domain-specific languages flourish, and powerful abstractions hide complexity. However, none of those developments comes close to the ubiquitous support for programming parallel hardware that is required to ensure that IT's effect on society over the next two decades will be as stunning as it has been over the last half-century.

For those reasons, the Committee on Sustaining Growth in Computing Performance recommends that our nation place a much greater emphasis on IT and computer-science research and development focused on improvements and innovations in parallel processing, and on making the transition to computing centered on parallelism. The following should have high priority:

- Algorithms that can exploit parallel processing;
- New computing "stacks" (applications, programming languages, compilers, runtime/virtual machines, operating systems, and architectures) that execute parallel rather than sequential programs and that effectively manage software parallelism, hardware parallelism, power, memory, and other resources;

- Portable programming models that allow expert and typical programmers to express parallelism easily and allow software to be efficiently reused on multiple generations of evolving hardware;
- Parallel-computing architectures driven by applications, including enhancements of chip multiprocessors, conventional data-parallel architectures, application-specific architectures, and radically different architectures;
- Open interface standards for parallel programming systems that promote cooperation and innovation to accelerate the transition to practical parallel computing systems; and
- Engineering and computer-science educational programs that incorporate an increased emphasis on parallelism and use a variety of methods and approaches to better prepare students for the types of computing resources that they will encounter in their careers.

Although all of those areas are important, fundamental power and energy constraints mean that even the best efforts might not yield a complete solution. Parallel computing systems will grow in performance over the long term only if they can become more power-efficient. Therefore, in addition to a focus on parallel processing, we need research and development on much more power-efficient computing systems at all levels of technology, including devices, hardware architecture, and software systems.

Summary

The end of dramatic exponential growth in single-processor performance marks the end of the dominance of the single microprocessor in computing. The era of sequential computing must give way to a new era in which parallelism is at the forefront. Although important scientific and engineering challenges lie ahead, this is an opportune time for innovation in programming systems and computing architectures. We have already begun to see diversity in computer designs to optimize for such considerations as power and throughput. The next generation of discoveries is likely to require advances at both the hardware and software levels of computing systems.

There is no guarantee that we can make parallel computing as common and easy to use as yesterday's sequential single-processor computer systems, but unless we aggressively pursue efforts suggested by the recommendations below, it will be "game over" for growth in computing performance. If parallel programming and related software efforts fail to become widespread, the development of exciting new applications that drive the computer industry will stall; if such innovation stalls, many other parts of the economy will follow suit.

This report of the Committee on Sustaining Growth in Computing Performance describes the factors that have led to the future limitations on growth for single processors based on complementary metal oxide semiconductor (CMOS) technology. The recommendations that follow are aimed at supporting and focusing research, development, and education in parallel computing, architectures, and power to sustain growth in computer performance and enjoy the next level of benefits to society.

SOCIETAL DEPENDENCE ON GROWTH IN COMPUTING PERFORMANCE

Information technology (IT) has transformed how we work and live—and has the potential to continue to do so. IT helps to bring distant people together, coordinate disaster response, enhance economic productivity, enable new medical diagnoses and treatments, add new efficiencies to our economy, improve weather prediction and climate modeling, broaden educational access, strengthen national defense, advance science, and produce and deliver content for education and entertainment.

Those transformations have been made possible by sustained improvements in the performance of computers. We have been living in a world where the cost of information processing has been decreasing exponentially year after year. The term *Moore's law*, which originally referred to an empirical observation about the most economically favorable rate for industry to increase the number of transistors on a chip, has come to be associated, at least popularly, with the expectation that microprocessors will become faster, that communication bandwidth will increase, that storage will become less expensive, and, more broadly, that computers will become faster. Most notably, the performance of individual computer processors increased on the order of 10,000 times over the last 2 decades of the 20th century without substantial increases in cost or power consumption.

Although some might say that they do not want or need a faster computer, computer users as well as the computer industry have in reality become dependent on the continuation of that performance growth. U.S. leadership in IT depends in no small part on taking advantage of the leading edge of computing performance. The IT industry annually generates a trillion dollars of revenue and has even larger indirect effects throughout society. This huge economic engine depends on a sustained demand for IT products and services; use of these products and services in turn fuels demand for constantly improving performance. More broadly, virtually every sector of society—manufacturing, financial services, education, science, government, the military, entertainment, and so on—has become dependent on continued growth in computing performance to drive industrial productivity, increase efficiency, and enable innovation. The performance achievements have driven an implicit, pervasive expectation that future IT advances will occur as an inevitable continuation of the stunning advances that IT has experienced over the last half-century.

SUMMARY 7

Finding: The information technology sector itself and most other sectors of society—for example, manufacturing, financial and other services, science, engineering, education, defense and other government services, and entertainment—have grown dependent on continued growth in computing performance.

Software developers themselves have come to depend on that growth in performance in several important ways, including:

- Developing applications that were previously infeasible, such as real-time video chat;
- Adding visible features and ever more sophisticated interfaces to existing applications;
- Adding "hidden" (nonfunctional) value—such as improved security, reliability, and other trustworthiness features—without degrading the performance of existing functions;
- Using higher-level abstractions, programming languages, and systems that require more computing power but reduce development time and improve software quality by making the development of correct programs and the integration of components easier; and
- Anticipating performance improvements and creating innovative, computationally intensive applications even before the required performance is available at low cost.

The U.S. computing industry has been adept at taking advantage of increases in computing performance, allowing the United States to be a moving and therefore elusive target—innovating and improvising faster than anyone else. If computer capability improvements stall, the U.S. lead will erode, as will the associated industrial competitiveness and military advantage.

Another consequence of 5 decades of exponential growth in performance has been the rise and dominance of the general-purpose microprocessor that is the heart of all personal computers. The dominance of the general-purpose microprocessor has stemmed from a virtuous cycle of (1) economies of scale wherein each generation of computers has been both faster and less expensive than the previous one, and (2) software correctness and performance portability—current software continues to run and to run faster on the new computers, and innovative applications can also run on them. The economies of scale have resulted from Moore's law scaling of transistor density along with innovative approaches to harnessing effectively all the new transistors that have become available. Portability has been preserved by keeping instruction sets compatible over many generations of microprocessors even as the underlying microprocessor technology saw substantial enhancements, allowing investments in software to be amortized over long periods.

The success of this virtuous cycle dampened interest in the development of alternative computer and programming models. Even though alternative architectures might have been technically superior (faster or more power-efficient) in specific domains, if they did not offer software compatibility they could not easily compete in the marketplace and were easily overtaken by the ever-improving general-purpose processors available at a relatively low cost.

CONSTRAINTS ON GROWTH IN SINGLE-PROCESSOR PERFORMANCE

By the 2000s, however, it had become apparent that processor performance growth was facing two major constraints. First, the ability to increase clock speeds has run up against power limits. The densest, highest-performance, and most power-efficient integrated circuits are constructed from CMOS technology. By 2004, the long-fruitful strategy of scaling down the size of CMOS circuits, reducing the supply voltage and increasing the clock rate was becoming infeasible. Since a chip's power consumption is proportional to the clock speed times the supply voltage squared, the inability to continue to lower the supply voltage halted the ability to increase the clock speed without increasing power dissipation. The resulting power consumption exceeded the few hundred watts per chip level that can practically be dissipated in a mass-market computing device as well as the practical limits for mobile, battery-powered devices. The ultimate consequence has been that growth in single-processor performance has stalled (or at least is being increased only marginally over time).

Second, efforts to improve the internal architecture of individual processors have seen diminishing returns. Many advances in the architecture of general-purpose sequential processors, such as deeper pipelines and speculative execution, have contributed to successful exploitation of increasing transistor densities. Today, however, there appears to be little opportunity to significantly increase performance by improving the internal structure of existing sequential processors. Figure S.1 graphically illustrates these trends and the slowdown in the growth of processor performance, clock speed, and power since around 2004. In contrast, it also shows the continued, exponential growth in the number of transistors per chip. The original Moore's law projection of increasing transistors per chip continues unabated even as performance has stalled. The 2009 edition of

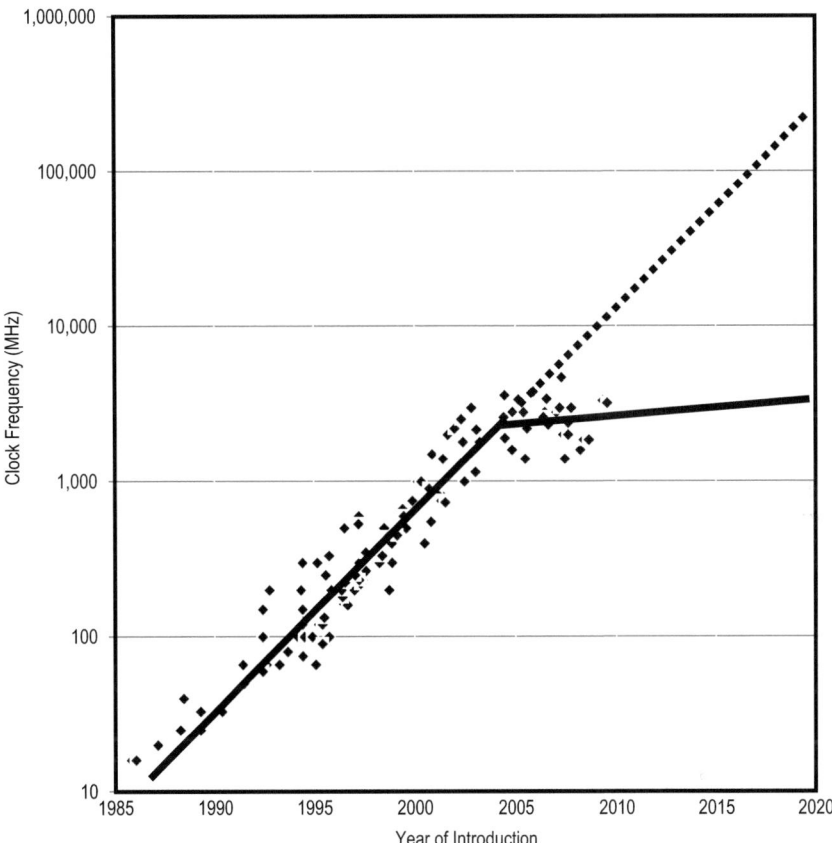

FIGURE S.1 Historical growth in single-processor performance and a forecast of processor performance to 2020, based on the ITRS roadmap. The dashed line represents expectations if single-processor performance had continued its historical trend. The vertical scale is logarithmic. A break in the growth rate at around 2004 can be seen. Before 2004, processor performance was growing by a factor of about 100 per decade; since 2004, processor performance has been growing and is forecasted to grow by a factor of only about 2 per decade. An expectation gap is apparent. In 2010, this expectation gap for single-processor performance is about a factor of 10; by 2020, it will have grown to a factor of 100. Most sectors of the economy and society implicitly or explicitly expect computing to deliver steady, exponentially increasing performance, but as this graph illustrates, traditional single-processor computing systems will not match expectations. Note that this graph plots processor clock rate as the measure of processor performance. Other processor design choices impact processor performance, but clock rate is a dominant processor performance determinant.

the International Technology Roadmap for Semiconductors (ITRS)[1] predicts the continued growth of transistors/chips for the next decade, but it will probably not be possible to continue to increase the transistor density (number of transistors per unit area) of CMOS chips at the current pace beyond the next 10 to 15 years. Figure S.1 shows the historical growth in single-processor performance and a forecast of processor performance to 2020 based on the ITRS roadmap. A dashed line represents what could have been expected if single-processor performance had continued on its historical trend. By 2020, however, a large "expectation gap" is apparent for single processors. This report explores the implications of that gap and offers a way to begin to bridging it.

Finding: After many decades of dramatic exponential growth, single-processor performance is increasing at a much lower rate, and this situation is not expected to improve in the foreseeable future.

Energy and power constraints play an important—and growing—role in computing performance. Computer systems require energy to operate and, as with any device, the more energy needed, the more expensive the system is to operate and maintain. Moreover, all the energy consumed by the system ends up as heat, which must be removed. Even when new parallel models and solutions are found, most future computing systems' performance will be limited by power or energy in ways that the computer industry and researchers have not had to deal with thus far. For example, the benefits of replacing a single, highly complex processor with increasing numbers of simpler processors will eventually reach a limit when further simplification costs more in performance than it saves in power. Constraints due to power are thus inevitable for systems ranging from hand-held devices to the largest computing data centers even as the transition is made to parallel systems.

Even with success in sidestepping the limits on single-processor performance, total energy consumption will remain an important concern, and growth in performance will become limited by power consumption within a decade. The total energy consumed by computing systems is already substantial and continues to grow rapidly in the United States and around the world. As is the case in other sectors of the economy, the total energy consumed by computing will come under increasing pressure.

---

[1]ITRS, 2009, ITRS 2009 Edition, available online at http://www.itrs.net/links/2009ITRS/Home2009.htm.

Finding: The growth in the performance of computing systems—even if they are multiple-processor parallel systems—will become limited by power consumption within a decade.

In short, the single processor and the sequential programming model that have dominated computing since its birth in the 1940s, will no longer be sufficient to deliver the continued growth in performance needed to facilitate future IT advances.[2] Moreover, it is an open question whether power and energy will be showstoppers or just significant constraints. Although these issues pose major technical challenges, they will also drive considerable innovation in computing by forcing a rethinking of the von Neumann model that has prevailed since the 1940s.

PARALLELISM AS A SOLUTION

Future growth in computing performance will have to come from parallelism. Most software developers today think and program by using a *sequential programming model* to create software for single general-purpose microprocessors. The microprocessor industry has already begun to deliver parallel hardware in mainstream products with chip multi-processors (CMPs—sometimes referred to as multicore), an approach that places new burdens on software developers to build applications to take advantage of multiple, distinct cores. Although developers have found reasonable ways to use two or even four cores effectively by running independent tasks on each one, they have not, for the most part, parallelized individual tasks in such a way as to make full use of the available computational capacity. Moreover, if industry continues to follow the same trends, they will soon be delivering chips with hundreds of cores. Harnessing these cores will require new techniques for parallel computing, including breakthroughs in software models, languages, and tools. Developers of both hardware and software will need to focus more attention on overall system performance, likely at the expense of time to market and the efficiency of the virtuous cycle described previously.

Of course, the computer science and engineering communities have been working for decades on the hard problems associated with parallelism. For example, high-performance computing for science and engineering applications has depended on particular parallel-programming techniques such as Message Passing Interface (MPI). In other cases,

---

[2]Of course, computing performance encompasses more than intrinsic CPU speed, but CPU performance has historically driven everything else: input/output, memory sizes and speeds, buses and interconnects, networks, and so on. If continued growth in CPU performance is threatened, so are the rest.

domain-specific languages and abstractions such as MapReduce have provided interfaces with behind-the-scenes parallelism and well-chosen abstractions developed by experts, technologies that hide the complexity of parallel programming from application developers. Those efforts have typically involved a small cadre of programmers with highly specialized training in parallel programming working on relatively narrow types of computing problems. None of this work has, however, come close to enabling widespread use of parallel programming for a wide array of computing problems.

Encouragingly, a few research universities, including MIT, the University of Washington, the University of California, Berkeley, and others have launched or revived research programs in parallelism, and the topic has also seen a renewed focus in industry at companies such as NVIDIA. However, these initial investments are not commensurate with the magnitude of the technical challenges or the stakes. Moreover, history shows that technology advances of this sort often require a decade or more. The results of such research are already needed today to sustain historical trends in computing performance, which makes us already a decade behind. Even with concerted investment, there is no guarantee that widely applicable solutions will be found. If they cannot be, we need to know that as soon as possible so that we can seek other avenues for progress.

Finding: There is no known alternative to parallel systems for sustaining growth in computing performance; however, no compelling programming paradigms for general parallel systems have yet emerged.

RECOMMENDATIONS

The committee's findings outline a set of serious challenges that affect not only the computing industry but also the many sectors of society that now depend on advances in IT and computation, and they suggest national and global economic repercussions. At the same time, the crisis in computing performance has pointed the way to new opportunities for innovation in diverse software and hardware infrastructures that excel in metrics other than single-chip processing performance, such as low power consumption and aggregate delivery of throughput cycles. There are opportunities for major changes in system architectures, and extensive investment in whole-system research is needed to lay the foundation of the computing environment for the next generation.

The committee's recommendations are broadly aimed at federal research agencies, the computing and information technology industry, and educators and fall into two categories. The first is research. The best science and engineering minds must be brought to bear on the challenges. The second is practice and education. Better practice in the development of computer hardware and software *today* will provide a foundation for future performance gains. Education will enable the emerging generation of technical experts to understand different and in some cases not-yet-developed parallel models of thinking about IT, computation, and software.

Recommendations for Research

The committee urges investment in several crosscutting areas of research, including algorithms, broadly usable parallel programming methods, rethinking the canonical computing stack, parallel architectures, and power efficiency.

Recommendation: Invest in research in and development of algorithms that can exploit parallel processing.

Today, relatively little software is explicitly parallel. To obtain the desired performance, it will be necessary for many more—if not most—software designers to grapple with parallelism. For some applications, they may still be able to write sequential programs, leaving it to compilers and other software tools to extract the parallelism in the underlying algorithms. For more complex applications, it may be necessary for programmers to write explicitly parallel programs. Parallel approaches are already used in some applications when there is no viable alternative. The committee believes that careful attention to parallelism will become the rule rather than the notable exception.

Recommendation: Invest in research in and development of programming methods that will enable efficient use of parallel systems not only by parallel-systems experts but also by typical programmers.

Many of today's programming models, languages, compilers, hypervisors (to manage virtual machines), and operating systems are targeted primarily at single-processor hardware. In the future, these layers will need to target, optimize programs for, and be optimized themselves for explicitly parallel hardware. The intellectual keystone of the endeavor is rethinking programming models so that programmers can express application parallelism naturally. The idea is to allow parallel software to be developed for diverse systems rather than specific configurations, and to have system software deal with balancing computation and minimizing communication among multiple computational units. The situation is reminiscent of the late 1970s, when programming models and tools were not up to the task of building substantially more complex software. Better programming models—such as structured programming in the 1970s, object orientation in the 1980s, and managed programming languages in the 1990s—have made it possible to produce much more sophisticated software. Analogous advances in the form of better tools and additional training will be needed to increase programmer productivity for parallel systems.

A key breakthrough would be the ability to express application parallelism in such ways that an application will run faster as more cores are added. The most prevalent parallel-programming languages do not provide this performance portability. A related question is what to do with the enormous body of legacy sequential code, which will be able to realize substantial performance improvements only if it can be parallelized. Experience has shown that parallelizing sequential code or highly sequential algorithms effectively is exceedingly difficult in general. Writing software that expresses the type of parallelism required to exploit chip multiprocessor hardware requires new software engineering processes and tools, including new programming languages that ease the expression of parallelism and a new software stack that can exploit and map the parallelism to hardware that is diverse and evolving. It will also require training programmers to solve their problems with parallel computational thinking.

The models themselves may or may not be explicitly parallel; it is an open question whether or when most programmers should be exposed to explicit parallelism. A single, universal programming model may or may not exist, and so multiple models—including some that are domain-specific—should be explored. Additional research is needed in the development of new libraries and new programming languages with appropriate compilation and runtime support that embody the new programming models. It seems reasonable to expect that some programming models, libraries, and languages will be suited for a broad base of skilled but not superstar programmers. They may even appear on the surface to be sequential or declarative. Others, however, will target efficiency, seeking the highest performance for critical subsystems that are to be extensively reused, and thus be intended for a smaller set of expert programmers.

Another focus for research should be system software for highly parallel systems. Although operating systems of today can handle some modest parallelism, future systems will include many more processors whose allocation, load balancing, and data communication and synchronization interactions will be difficult to handle well. Solving those problems will require a rethinking of how computation and communication resources are viewed much as demands for increased memory size led to the introduction of virtual memory a half-century ago.

Recommendation: Focus long-term efforts on rethinking of the canonical computing "stack"—applications, programming language, compiler, runtime, virtual machine, operating system, hypervisor, and architecture—in light of parallelism and resource-management challenges.

Computer scientists and engineers typically manage complexity by separating interface from implementation. In conventional computer systems, that is done recursively to form a computing stack: applications, programming language, compiler, runtime, virtual machine, operating system, hypervisor, and architecture. It is unclear whether today's conventional stack provides the right framework to support parallelism and manage resources. The structure and elements of the stack itself should be a focus of long-term research exploration.

Recommendation: Invest in research on and development of parallel architectures driven by applications, including enhancements of chip multiprocessor systems and conventional data-parallel architectures, cost-effective designs for application-specific architectures, and support for radically different approaches.

In addition to innovation and advancements in parallel programming models and systems, advances in architecture and hardware will play an important role. One path forward is to continue to refine the chip multiprocessors (CMPs) and associated architectural approaches. Are today's CMP approaches suitable for designing most computers? The current CMP architecture has the advantage of maintaining compatibility with existing software, the heart of the architectural franchise that keeps companies investing heavily. But CMP architectures bring their own challenges. Will large numbers of cores work in most computer deployments, such as on desktops and even in mobile phones? How can cores be harnessed together temporarily, in an automated or semiautomated fashion, to overcome sequential bottlenecks? What mechanisms and policies will best exploit locality (keeping data stored close to other data that might be needed at the same time or for particular computations and saving on the power needed to move data around) so as to avoid communications bottlenecks? How should synchronization and scheduling be handled? How should challenges associated with power and energy be addressed? What do the new architectures mean for such system-level features as reliability and security?

Is using homogeneous processors in CMP architectures the best approach, or will computer architectures that include multiple but heterogeneous cores—some of which may be more capable than others or even use different instruction-set architectures—be more effective? Special-purpose processors that have long exploited parallelism, notably graphics processing units (GPUs) and digital signal processor (DSP) hardware, have been successfully deployed in important segments of the market. Are there other important niches like those filled by GPUs and DSPs? Alternatively, will computing cores support more graphics and GPUs support more general-purpose programs, so that the difference between the two will blur?

Perhaps some entirely new architectural approach will prove more successful. If systems with CMP architectures cannot be effectively programmed, an alternative will be needed. Work in this general area could eschew conventional cores. It can view the chip as a *tabula rasa* of billions of transistors, which translates to hundreds of functional units; the effective organization of those units into a programmable architecture is an open question. Exploratory computing systems based on field-programmable gate arrays (FPGAs) are a step in this direction, but continued innovation is needed to develop programming systems that can harness the potential parallelism of FPGAs.

Another place where fundamentally different approaches may be needed is alternatives to CMOS. There are many advantages to sticking with today's silicon-based CMOS technology, which has proved remarkably scalable over many generations of microprocessors and around which an enormous industrial and experience base has been established. However, it will also be essential to invest in new computation substrates whose underlying power efficiency promises to be fundamentally better than that of silicon-based CMOS circuits. Computing has benefited in the past from order-of-magnitude performance improvements in power consumption in the progression from vacuum tubes to discrete bipolar transistors to integrated circuits first based on bipolar transistors, then on N-type metal oxide semiconductors (NMOS) and now on CMOS. No alternative is near commercial availability yet, although some show potential.

In the best case, investment will yield devices and manufacturing methods—as yet unforeseen—that will dramatically surpass the CMOS IC. In the worst case, no new technology will emerge to help solve the problems. That uncertainty argues for investment in multiple approaches as soon as possible, and computer system designers would be well advised not to expect one of the new devices to appear in time to obviate the development of new, parallel architectures built on the proven CMOS technology. Better performance is needed immediately. Society cannot wait the decade or two that it would take to identify, refine, and apply a new technology that may or may not even be on the horizon now. Moreover, even if a groundbreaking new technology were discovered, the investment in parallelism would not be wasted, in that advances in parallelism would probably exploit the new technology as well.

Recommendation: Invest in research and development to make computer systems more power-efficient at all levels of the system, including software, application-specific approaches, and alternative devices. Such efforts should address ways in which software and system architectures can improve power efficiency, such as by exploiting locality and the use of domain-specific execution units. R&D should also be aimed at making logic gates more power-efficient. Such efforts should address alternative physical devices beyond incremental improvements in today's CMOS circuits.

Because computing systems are increasingly limited by energy consumption and power dissipation, it is essential to invest in research and development to make computing systems more power-efficient. Exploiting parallelism alone cannot ensure continued growth in computer performance. There are numerous potential avenues for investigation into better power efficiency, some of which require sustained attention to known engineering issues and others of which require research. These include:

- Redesign the delivery of power to and removal of heat from computing systems for increased efficiency. Design and deploy systems in which the absolute maximum fraction of power is used to do the computing and less is used in routing power to the system and removing heat from the system. New voluntary or mandatory standards (including ones that set ever-more-aggressive targets) might provide useful incentives for the development and use of better techniques.
- Develop alternatives to the general-purpose processor that exploit locality.
- Develop domain-specific or application-specific processors analogous to GPUs and DSPs that provide better performance and power-consumption characteristics than do general-purpose processors for other specific application domains.
- Investigate possible new, lower-power device technology beyond CMOS.

Additional research should focus on system designs and software configurations that reduce power consumption, for example, reducing power consumption when resources are idle, mapping applications to domain-specific and heterogeneous hardware units, and limiting the amount of communication among disparate hardware units.

Although the shift toward CMPs will allow industry to continue for some time to scale the performance of CMPs based on general-purpose processors, general-purpose CMPs will eventually reach their own limits. CMP designers can trade off single-thread performance of individual processors against lower energy dissipation per instruction, thus allowing more instructions by multiple processors while the same amount of energy is dissipated by the chip. However, that is possible only within a limited range of energy performance. Beyond some limit, lowering energy per instruction by processor simplification can lead to degradation in overall CMP performance because processor performance starts to decrease faster than energy per instruction. When that occurs, new approaches will be needed to create more energy-efficient computers.

It may be that general-purpose CMPs will prove not to be a solution in the long run and that we will need to create more application-optimized processing units. Tuning hardware and software toward a specific type of application allows a much more energy-efficient solution. However, the current design trend is away from building customized solutions, because increasing design complexity has caused the nonrecurring engineering costs for designing the chips to grow rapidly. High costs limit the range of potential market segments to the few that have volume high enough to justify the initial engineering investment. A shift to more application-optimized computing systems, if necessary, demands a new approach to design that would allow application-specific chips to be created at reasonable cost.

Recommendations for Practice and Education

Implementing the research agenda proposed here, although crucial for progress, will take time. Meanwhile, society has an immediate and pressing need to use current and emerging CMP systems effectively. To that end, the committee offers three recommendations related to current development and engineering practices and educational opportunities.

Recommendation: To promote cooperation and innovation by sharing, encourage development of open interface standards for parallel programming rather than proliferating proprietary programming environments.

Private-sector firms are often incentivized to create proprietary interfaces and implementations to establish a competitive advantage. However, a lack of standardization can impede progress inasmuch as the presence of many incompatible approaches allows none to achieve the benefits of wide adoption and reuse—a major reason that industry participates in standards efforts. The committee encourages the development of programming-interface standards that can facilitate wide adoption of parallel programming even as they foster competition in other matters.

Recommendation: Invest in the development of tools and methods to transform legacy applications to parallel systems.

Whatever long-term success is achieved in the effective use of parallel systems from rethinking algorithms and developing new programming methods will probably come at the expense of the backward-platform and cross-platform compatibility that has been an economic cornerstone of IT for decades. To salvage value from the nation's current, substantial IT investment, we should seek ways to bring sequential programs into the parallel world. On the one hand, there are probably no "silver bullets" that enable automatic transformation. On the other hand, it is prohibitively expensive to rewrite many applications. The committee urges industry and academe to develop "power tools" for experts that can help them to migrate legacy code to tomorrow's parallel computers. In addition, emphasis should be placed on tools and strategies to enhance code creation, maintenance, verification, and adaptation of parallel programs.

Recommendation: Incorporate in computer science education an increased emphasis on parallelism, and use a variety of methods and approaches to better prepare students for the types of computing resources that they will encounter in their careers.

Who will develop the parallel software of the future? To sustain IT innovation, we will need a workforce that is adept in writing parallel applications that run well on parallel hardware, in creating parallel software systems, and in designing parallel hardware.

Both undergraduate and graduate students in computer science, as well as in other fields that make intensive use of computing, will need to be educated in parallel programming. The engineering, science, and computer-science curriculum at both the undergraduate and graduate levels should begin to incorporate an emphasis on parallel computational thinking, parallel algorithms, and parallel programming. With respect to the computer-science curriculum, because no general-purpose paradigm has emerged, universities should teach diverse parallel-programming languages, abstractions, and approaches until effective ways of teaching and programming emerge. The necessary shape of the needed changes will not be clear until some reasonably general parallel-programming methods have been devised and shown to be promising.

Related to this is improving the ability of the programming workforce to cope with the new challenges of parallelism. This will involve retraining today's programmers and also developing new models and abstractions to make parallel programming more accessible to typically skilled programmers.

CONCLUDING REMARKS

There is no guarantee that we can make future parallel computing ubiquitous and as easy to use as yesterday's sequential computer, but unless we aggressively pursue efforts as suggested by the recommendations above, it will be game over for future growth in computing performance. This report describes the factors that have led to limitations of growth of single processors based on CMOS technology. The recommendations here are aimed at supporting and focusing research, development, and education in architectures, power, and parallel computing to sustain growth in computer performance and to permit society to enjoy the next level of benefits.

1

The Need for Continued Performance Growth

Information technology (IT) has become an integral part of modern society, affecting nearly every aspect of our lives, including education, medicine, government, business, entertainment, and social interactions. Innovations in IT have been fueled by a continuous and extraordinary increase in computer performance. By some metrics computer performance has improved by a factor of an average of 10 every 5 years over the past 2 decades. A sustained downshift in the rate of growth in computing performance would have considerable ramifications both economically and for society. The industries involved are responsible for about $1 trillion of annual revenue in the United States. That revenue has depended on a sustained demand for IT products and services that in turn has fueled demand for constantly improving performance. Indeed, U.S. leadership in IT depends in no small part on its driving and taking advantage of the leading edge of computing performance. Virtually every sector of society—manufacturing, financial services, education, science, government, military, entertainment, and so on—has become dependent on the continued growth in computing performance to drive new efficiencies and innovations. Moreover, all the current and foreseeable future applications rely on a huge software infrastructure, and the software infrastructure itself would have been impossible to develop with the more primitive software development and programming methods of the past. The principal force allowing better programming models, which emphasize programmer productivity over computing efficiency, has been the growth in computing performance. (Chapter 4 explores implications for software and programming in more detail.)

This chapter first considers the general question of why faster computers are important. It then examines four broad fields—science, defense and national security, consumer applications, and enterprise productivity—that have depended on and will continue to depend on sustained growth in computing performance. The fields discussed by no means constitute an exhaustive list,[1] but they are meant to illustrate how computing performance and its historic exponential growth have had vast effects on broad sectors of society and what the results of a slowdown in that growth would be.

WHY FASTER COMPUTERS ARE IMPORTANT

Computers can do only four things: they can move data from one place to another, they can create new data from old data via various arithmetic and logical operations, they can store data in and retrieve them from memories, and they can decide what to do next. Students studying computers or programming for the first time are often struck by the surprising intuition that, notwithstanding compelling appearance to the contrary, computers are extremely primitive machines, capable of performing only the most mind-numbingly banal tasks. The trick is that computers can perform those simple tasks extremely fast—in periods measured in billionths of a second—and they perform these tasks reliably and repeatably. Like a drop of water in the Grand Canyon, each operation may be simple and may in itself not accomplish much, but a lot of them (billions per second, in the case of computers) can get a lot done.

Over the last 60 years of computing history, computer buyers and users have essentially "voted with their wallets" by consistently paying more for faster computers, and computer makers have responded by pric-

---

[1] Health care is another field in which IT has substantial effects—in, for example, patient care, research and innovation, and administration. A recent National Research Council (NRC) report, although it does not focus specifically on computing performance, provides numerous examples of ways in which computation technology and IT are critical underpinnings of virtually every aspect of health care (NRC, 2009, Computational Technology for Effective Health Care: Immediate Steps and Strategic Directions, Washington, D.C.: The National Academies Press, available online at http://www.nap.edu/catalog.php?record_id=12572). Yet another critically important field that increasingly benefits from computation power is infrastructure. "Smart" infrastructure applications in urban planning, high-performance buildings, energy, traffic, and so on are of increasing importance. That is also the underlying theme of two of the articles in the February 2009 issue of Communications of the ACM (Tom Leighton, 2009, Improving performance on the Internet, Communications of the ACM 52(2): 44-51; and T.V. Raman, 2009, Toward 2^W: Beyond Web 2.0, Communications of the ACM 52(2): 52-59).

ing their systems accordingly: a high-end system may be, on the average, 10 percent faster and 30 percent more expensive than the next-best. That behavior has dovetailed perfectly with the underlying technology development in the computers—as ever-faster silicon technology has become available, faster and faster computers could be designed. It is the nature of the semiconductor manufacturing process that silicon chips coming off the fabrication line exhibit a range of speeds. Rather than discard the slower chips, the manufacturer simply charges less for them. Ever-rising performance has been the wellspring of the entire computer industry. Meanwhile, the improving economics of ever-larger shipment volumes have driven overall system costs down, reinforcing a virtuous spiral[2] by making computer systems available to lower-price, larger-unit-volume markets.

For their part, computer buyers demand ever-faster computers in part because they believe that using faster machines confers on them an advantage in the marketplace in which they compete.[3] Applications that run on a particular generation of computing system may be impractical or not run at all on a system that is only one-tenth as fast, and this encourages hardware replacements for performance every 3-5 years. That trend has also encouraged buyers to place a premium on fast new computer systems because buying fast systems will forestall system obsolescence as long as possible. Traditionally, software providers have shown a tendency to use exponentially more storage space and central processing unit (CPU) cycles to attain linearly more performance; a tradeoff commonly referred to as bloat. Reducing bloat is another way in which future system improvements may be possible. The need for periodic replacements exists whether the performance is taking place on the desktop or in the "cloud"

---

[2] A small number of chips are fast, and many more are slower. That is how a range of products is produced that in total provide profits and, ultimately, funding for the next generation of technology. The semiconductor industry is nearing a point where extreme ultraviolet (EUV) light sources—or other expensive, exotic alternatives—will be needed to continue the lithography-based steps in manufacturing. There are a few more techniques left to implement before EUV is required, but they are increasingly expensive to use in manufacturing, and they are driving costs substantially higher. The future scenario that this implies is not only that very few companies will be able to manufacture chips with the smallest feature sizes but also that only very high-volume products will be able to justify the cost of using the latest generation of technology.

[3] For scientific researchers, faster computers allow larger or more important questions to be pursued or more accurate answers to be obtained; office workers can model, communicate, store, retrieve, and search their data more productively; engineers can design buildings, bridges, materials, chemicals, and other devices more quickly and safely; and manufacturers can automate various parts of their assembly processes and delivery methods more cost-effectively. In fact, the increasing amounts of data that are generated, stored, indexed, and retrieved require continued performance improvements. See Box 1.1 for more on data as a performance driver.

**BOX 1.1
Growth of Stored and Retrievable Data**

The quantity of information and data that is stored in a digital format has been growing at an exponential rate that exceeds even the historical rate of growth in computing performance, which is the focus of this report. Data are of value only if they can be analyzed to produce useful information that can be retrieved when needed. Hence, the growth in stored information is another reason for the need to sustain substantial growth in computing performance.

As the types and formats of information that is stored in digital form continue to increase, they drive the rapid growth in stored data. Only a few decades ago, the primary data types stored in IT systems were text and numerical data. But images of increasing resolution, audio streams, and video have all become important types of data stored digitally and then indexed, searched, and retrieved by computing systems.

The growth of stored information is occurring at the personal, enterprise, national, and global levels. On the personal level, the expanding use of e-mail, text messaging, Web logs, and so on is adding to stored text. Digital cameras have enabled people to store many more images in their personal computers and data centers than they ever would have considered with traditional film cameras. Video cameras and audio recorders add yet more data that are stored and then must be indexed and searched. Embedding those devices into the ubiquitous cell phone means that people can and do take photos and movies of events that would previously not have been recorded.

At the global level, the amount of information on the Internet continues to increase dramatically. As static Web pages give way to interactive pages and social-networking sites support video, the amount of stored and searchable data continues its explosive growth. Storage technology has enabled this growth by reducing the cost of storage by a rate even greater than that of the growth in processor performance.

The challenge is to match the growth in stored information with the computational capability to index, search, and retrieve relevant information. Today, there are not sufficiently powerful computing systems to process effectively all the images and video streams being stored. Satellite cameras and other remote sensing devices typically collect much more data than can be examined for useful information or important events.

Considerably more progress is needed to achieve the vision described by Vannevar Bush in his 1945 paper about a MEMEX device that would collect and make available to users all the information relevant to their life and work.[1]

---

[1] Vannevar Bush, 1945, "As we may think," Atlantic Magazine, July 1945, available online at http://www.theatlantic.com/magazine/archive/1969/12/as-we-may-think/3881/.

in a Web-based service, although the pace of hardware replacement may vary in the cloud.

All else being equal, faster computers are better computers.[4] The unprecedented evolution of computers since 1980 exhibits an essentially exponential speedup that spans 4 orders of magnitude in performance for the same (or lower) price. No other engineered system in human history has ever achieved that rate of improvement; small wonder that our intuitions are ill-tuned to perceive its significance. Whole fields of human endeavors have been transformed as computer system capability has ascended through various threshold performance values.[5] The impact of computer technology is so widespread that it is nearly impossible to overstate its importance.

Faster computers create not just the ability to do old things faster but the ability to do new things that were not feasible at all before.[6] Fast computers have enabled cell phones, MP3 players, and global positioning devices; Internet search engines and worldwide online auctions; MRI and CT scanners; and handheld PDAs and wireless networks. In many cases, those achievements were not predicted, nor were computers designed specifically to cause the breakthroughs. There is no overarching roadmap for where faster computer technology will take us—each new achievement opens doors to developments that we had not even conceived. We should assume that this pattern will continue as computer systems

---

[4]See Box 1.2 for a discussion of why this is true even though desktop computers, for example, spend most of their time idle.

[5]The music business, for example, is almost entirely digital now, from the initial sound capture through mixing, processing, mastering, and distribution. Computer-based tricks that were once almost inconceivable are now commonplace, from subtly adjusting a singer's note to be more in tune with the instruments, to nudging the timing of one instrument relative to another. All keyboard instruments except acoustic pianos are now digital (computer-based) and not only can render very accurate imitations of existing instruments but also can alter them in real time in a dizzying variety of ways. It has even become possible to isolate a single note from a chord and alter it, a trick that had long been thought impossible. Similarly, modern cars have dozens of microprocessors that run the engine more efficiently, minimize exhaust pollution, control the antilock braking system, control the security system, control the sound system, control the navigation system, control the airbags and seatbelt retractors, operate the cruise control, and handle other features. Over many years, the increasing capability of these embedded computer systems has allowed them to penetrate nearly every aspect of vehicles.

[6]Anyone who has played state-of-the-art video games will recognize the various ways in which game designers wielded the computational and graphics horsepower of a new computer system for extra realism in a game's features, screen resolution, frame rate, scope of the "theater of combat," and so on.

> **BOX 1.2**
> **Why Do I Need More Performance When
> My Computer Is Idle Most of the Time?**
>
> When computers find themselves with nothing to do, by default they run operating-system code known as the idle loop. The idle loop is like the cell-phone parking lot at an airport, where your spouse sits waiting to pick you up when you arrive and call him or her. It may seem surprising or incongruous that nearly all the computing cycles ever executed by computers have been wasted in the idle loop, but it is true. If we have "wasted" virtually all the computing horsepower available since the beginning of the computer age, why should we worry about a potential threat to increased performance in the future? Is there any point in making machinery execute the idle loop even faster? In fact, there is. The reason has as much to do with humans as it does with the computing machines that they design.
>
> Consider the automobile. The average internal-combustion vehicle has a six-cylinder engine capable of a peak output of around 200 horsepower. Many aspects of the engine and drivetrain reflect that peak horsepower: when you press the pedal to the floor while passing or entering a highway, you expect the vehicle to deliver that peak horsepower to the wheels, and you would be quite unhappy if various parts of the car were to leave the vehicle instead, unable to handle the load. But if you drive efficiently, over several years of driving, what fraction of the time is spent under that peak load condition? For most people, the answer is approximately zero. It only takes about 20 horsepower to keep a passenger car at highway speeds under nominal conditions, so you end up paying for a lot more horsepower than you use.
>
> But if all you had at your driving disposal was a 20-horsepower power plant (essentially, a golf cart), you would soon tire of driving the vehicle because you would recognize that energy efficiency is great but not everything; that annoying all the other drivers as you slowly, painfully accelerate from an on-ramp gets old quickly; and that your own time is valuable to you as well. In effect, we all

get faster yet.[7] There is no reason to think that it will not continue as long as computers continue to improve. What has changed—and will be described in detail in later chapters—is how we achieve faster computers. In short, power dissipation can no longer be dealt with independently of performance (see Chapter 3). Moreover, although computing performance has many components (see Chapter 2), a touchstone in this report will be computer speed; as described in Box 1.3, speed can be traded for almost any other sort of functionality that one might want.

---

[7]Some of the breakthroughs were not solely performance-driven—some depend on a particular performance at a particular cost. But cost and performance are closely related, and performance can be traded for lower cost if desired.

accept a compromise that results in a system that is overdesigned for the common case because we care about the uncommon case and are willing to pay for the resulting inefficiency.

In a computing system, although you may know that the system is spending almost all its time doing nothing, that fact pales in comparison with how you feel when you ask the system to do something in real time and must wait for it to accomplish that task. For instance, when you click on an attachment or a file and are waiting for the associated application to open (assuming that it is not already open), every second drags.[1] At that moment, all you want is a faster system, regardless of what the machine is doing when you are not there. And for the same reason that a car's power plant and drivetrain are overdesigned for their normal use, your computing system will end up sporting clock frequencies, bus speeds, cache sizes, and memory capacity that will combine to yield a computing experience to you, the user, that is statistically rather rare but about which you care very much.

The idle-loop effect is much less pronounced in dedicated environments—such as servers and cloud computing, scientific supercomputers, and some embedded applications—than it is on personal desktop computers. Servers and supercomputers can never go fast enough, however—there is no real limit to the demand for higher performance in them. Some embedded applications, such as the engine computer in a car, will idle for a considerable fraction of their existence, but they must remain fast enough to handle the worst-case computational demands of the engine and the driver. Other embedded applications may run at a substantial fraction of peak capacity, depending on the workload and the system organization.

---

[1] It is worth noting that the interval between clicking on most e-mail attachments and successful opening of their corresponding applications is not so much a function of the CPU's performance as it is of disk speed, memory capacity, and input/output interconnect bandwidth.

Finding: The information technology sector itself and most other sectors of society—for example, manufacturing, financial and other services, science, engineering, education, defense and other government services, and entertainment—have become dependent on continued growth in computing performance.

The rest of this chapter describes a sampling of fields in which computing performance has been critical and in which a slowing of the growth of computing performance would have serious adverse repercussions. We focus first on high-performance computing and computing performance in the sciences. Threats to growth in computing performance will be felt there first, before inevitably extending to other types of computing.

> **BOX 1.3**
> **Computing Performance Is Fungible**
>
> Computing speed can be traded for almost any other feature that one might want. In this sense, computing-system performance is fungible, and that is what gives it such importance. Workloads that are at or near the absolute capacity of a computing system tend to get all the publicity—for every new computing-system generation, the marketing holy grail is a "killer app" (software application), some new software application that was previously infeasible, now runs adequately, and is so desirable that buyers will replace their existing systems just to buy whatever hardware is fast enough to run it. The VisiCalc spreadsheet program on the Apple II was the canonical killer app; it appeared at the dawn of the personal-computing era and was so compelling that many people bought computers just to run it. It has been at least a decade since anything like a killer app appeared, at least outside the vocal but relatively small hard-core gaming community. The reality is that modern computing systems spend nearly all their time idle (see Box 1.2 for an explanation of why faster computers are needed despite that); thus, most systems have a substantial amount of excess computing capacity, which can be put to use in other ways.
>
> Performance can be traded for higher reliability: for example, the digital signal processor in a compact-disk player executes an elaborate error-detection-and-correction algorithm, and the more processing capability can be brought to bear on that problem, the more bumps and shocks the player can withstand before the errors become audible to the listener. Computational capacity can also be used to index mail and other data on a computer periodically in the background to make the next search faster. Database servers can take elaborate precautions to ensure high system dependability in the face of inevitable hardware-component failures. Spacecraft computers often incorporate three processors where one would suffice for performance; the outputs of all three processors are compared via a voting scheme that detects if one of the three machines has failed. In effect, three processors' worth of performance is reduced to one processor's performance in exchange for improved system dependability. Performance can be used in the service of other goals, as well. Files on a hard drive can be compressed, and this trades computing effort and time for better effective drive capacity. Files that are sent across a network or across the Internet use far less bandwidth and arrive at their destination faster when they are compressed. Likewise, files can be encrypted in much the same way to keep their contents private while in transit.

THE IMPORTANCE OF COMPUTING PERFORMANCE FOR THE SCIENCES

Computing has become a critical component of most sciences and complements the traditional roles of theory and experimentation.[8] Theoretical models may be tested by implementing them in software, evaluating them through simulation, and comparing their results with known experimental results. Computational techniques are critical when experimentation is too expensive, too dangerous, or simply impossible. Examples include understanding the behavior of the universe after the Big Bang, the life cycle of stars, the structure of proteins, functions of living cells, genetics, and the behavior of subatomic particles. Computation is used for science and engineering problems that affect nearly every aspect of our daily lives, including the design of bridges, buildings, electronic devices, aircraft, medications, soft-drink containers, potato chips, and soap bubbles. Computation makes automobiles safer, more aerodynamic, and more energy-efficient. Extremely large computations are done to understand economics, national security, and climate change, and some of these computations are used in setting public policy. For example, hundreds of millions of processor hours are devoted to understanding and predicting climate change—one purpose of which is to inform the setting of international carbon-emission standards.

In many cases, what scientists and engineers can accomplish is limited by the performance of computing systems. With faster systems, they could simulate critical details—such as clouds in a climate model or mechanics, chemistry, and fluid dynamics in the human body—and they could run larger suites of computations that would improve confidence in the results of simulations and increase the range of scientific exploration.

Two themes common to many computational science and engineering disciplines are driving increases in computational capability. The first is an increased desire to support multiphysics or coupled simulations, such as adding chemical models to simulations that involve fluid-dynamics simulations or structural simulations. Multiphysics simulations are necessary for understanding complex real-world systems, such as the climate, the human body, nuclear weapons, and energy production. Imagine, for example, a model of the human body in which one could experiment with the addition of new chemicals (medicines to change blood pressure), changing structures (artificial organs or prosthetic devices), or effects of radiation. Many scientific fields are ripe for multiphysics simulations

---

[8]See an NRC report for one relatively recent take on computing and the sciences (NRC, 2008, The Potential Impact of High-End Capability Computing on Four Illustrative Fields of Science and Engineering, Washington, D.C.: The National Academies Press, available online at http://www.nap.edu/catalog.php?record_id=12451).

because the individual components are understood well enough and are represented by a particular model and instantiation within a given code base. The next step is to take two or more such code bases and couple them in such a way that each communicates with the others. Climate modeling, for example, is well along that path toward deploying coupled models, but the approach is still emerging in some other science domains.

The second crosscutting theme in the demand for increased computing performance is the need to improve confidence in simulations to make computation truly predictive. At one level, this may involve running multiple simulations and comparing results with different initial conditions, parameterizations, simulations at higher space or time resolutions or numerical precision, models, levels of detail, or implementations. In some fields, sophisticated "uncertainty quantification" techniques are built into application codes by using statistical models of uncertainty, redundant calculations, or other approaches. In any of those cases, the techniques to reduce uncertainty increase the demand for computing performance substantially.[9]

High-Energy Physics, Nuclear Physics, and Astrophysics

The basic sciences, including physics, also rely heavily on high-end computing to solve some of the most challenging questions involving phenomena that are too large, too small, or too far away to study directly. The report of the 2008 Department of Energy (DOE) workshop on Scientific Grand Challenges: Challenges for Understanding the Quantum Universe and the Role of Computing at the Extreme Scale summarizes the computational gap: "To date, the computational capacity has barely been able to keep up with the experimental and theoretical research programs. There is considerable evidence that the gap between scientific aspiration and the availability of computing resource is now widening. . . ."[10] One of the examples involves understanding properties of dark matter and dark energy by analyzing datasets from digital sky surveys, a technique that has already been used to explain the behavior of the universe shortly

---

[9]In 2008 and 2009, the Department of Energy (DOE) held a series of workshops on computing and extreme scales in a variety of sciences. The workshop reports summarize some of the scientific challenges that require 1,000 times more computing than is available to the science community today. More information about these workshops and others is available online at DOE's Office of Advanced Scientific Computing Research website, http://www.er.doe.gov/ascr/WorkshopsConferences/WorkshopsConferences.html.

[10]DOE, 2009, Scientific Grand Challenges: Challenges for Understanding the Quantum Universe and the Role of Computing at the Extreme Scale, Workshop Report, Menlo Park, Cal., December 9-11, 2008, p. 2, available at http://www.er.doe.gov/ascr/ProgramDocuments/ProgDocs.html.

after the Big Bang and its continuing expansion. The new datasets are expected to be on the order of 100 petabytes (10^{17} bytes) in size and will be generated with new high-resolution telescopes that are on an exponential growth path in capability and data generation. High-resolution simulations of type Ia and type II supernova explosions will be used to calibrate their luminosity; the behavior of such explosions is of fundamental interest, and such observational data contribute to our understanding of the expansion of the universe. In addition, an improved understanding of supernovae yields a better understanding of turbulent combustion under conditions not achievable on Earth. Finally, one of the most computationally expensive problems in physics is aimed at revealing new physics beyond the standard model, described in the DOE report as "analogous to the development of atomic physics and quantum electrodynamics in the 20th century."[11]

In addition to the data analysis needed for scientific experiments and basic compute-intensive problems to refine theory, computation is critical to engineering one-of-a-kind scientific instruments, such as particle accelerators like the International Linear Collider and fusion reactors like ITER (which originally stood for International Thermonuclear Experimental Reactor). Computation is used to optimize the designs, save money in construction, and reduce the risk associated with these devices. Similarly, simulation can aid in the design of complex systems outside the realm of basic science, such as nuclear reactors, or in extending the life of existing reactor plants.

Chemistry, Materials Science, and Fluid Dynamics

A 2003 National Research Council report outlines several of the "grand challenges" in chemistry and chemical engineering, including two that explicitly require high-performance computing.[12] The first is to "understand and control how molecules react—over all time scales and the full range of molecular size"; this will require advances in predictive computational modeling of molecular motions, which will complement other experimental and theoretical work. The second is to "learn how to design and produce new substances, materials, and molecular devices with properties that can be predicted, tailored, and tuned before production"; this will also require advances in computing and has implications for commercial use of chemical and materials engineering in medicine,

---

[11]Ibid. at p. vi.
[12]NRC, 2003, Beyond the Molecular Frontier: Challenges for Chemistry and Chemical Engineering, Washington, D.C.: The National Academies Press, available online at http://www.nap.edu/catalog.php?record_id=10633.

energy and defense applications, and other fields. Advances in computing performance are necessary to increase length scales to allow modeling of multigranular samples, to increase time scales for fast chemical processes, and to improve confidence in simulation results by allowing first-principles calculations that can be used in their own right or to validate codes based on approximate models. Computational materials science contributes to the discovery of new materials. The materials are often the foundation of new industries; for example, understanding of semiconductors led to the electronics industry and understanding of magnetic materials contributed to data storage.

Chemistry and material science are keys to solving some of the most pressing problems facing society today. In energy research, for example, they are used to develop cleaner fuels, new materials for solar panels, better batteries, more efficient catalysts, and chemical processes for carbon capture and sequestration. In nuclear energy alone, simulation that combines materials, fluids, and structures may be used for safety assessments, design activities, cost, and risk reduction.[13] Fluid-dynamics simulations are used to make buildings, engines, planes, cars, and other devices more energy-efficient and to improve understanding of the processes, such as combustion, that are fundamental to the behavior of stars, weapons, and energy production. Those simulations vary widely among computational scales, but whether they are run on personal computers or on petascale systems, the value of additional performance is universal.

Biological Sciences

The use of large-scale computation in biology is perhaps most visible in genomics, in which enormous data-analysis problems were involved in computing and mapping the human genome. Human genomics has changed from a purely science-driven field to one with commercial and personal applications as new sequence-generation systems have become a commodity and been combined with computing and storage systems that are modest by today's standards. Companies will soon offer personalized genome calculation to the public. Genomics does not stop with the human genome, however, and is critical in analyzing and synthesizing microorganisms for fighting disease, developing better biofuels, and mitigating environmental effects. The goal is no longer to sequence a single species but to scoop organisms from a pond or ocean, from soil, or from deep

---

[13] Horst Simon, Thomas Zacharia, and Rick Stevens, 2007, Modeling and Simulation at the Exascale for the Energy and Environment, Report on the Advanced Scientific Computing Research Town Hall Meetings on Simulation and Modeling at the Exascale for Energy, Ecological Sustainability and Global Security (E3), Washington, D.C.: DOE, available online at http://www.er.doe.gov/ascr/ProgramDocuments/Docs/TownHall.pdf.

underground and analyze an entire host of organisms and compare them with other species to understand better what lives and why in particular environments.

At the macro level, reverse engineering of the human brain and simulating complete biologic systems from individual cells to the structures and fluids of a human are still enormous challenges that exceed our current reach in both understanding and computational capability. But progress on smaller versions of those problems shows that progress is possible.

One of the most successful kinds of computation in biology has been at the level of proteins and understanding their structure. For example, a group of biochemical researchers[14] are applying standard computer-industry technology (technology that was originally designed with and funded by profits from mundane consumer electronics items) to tackle the protein-folding problem at the heart of modern drug discovery and invention. This problem has eluded even the fastest computers because of its overwhelming scale and complexity. But several decades of Moore's law have now enabled computational machinery of such capability that the protein-folding problem is coming into range. With even faster hardware in the future, new treatment regimens tailored to individual patients may become feasible with far fewer side effects.

Climate-Change Science

In its 2007 report on climate change, the Intergovernmental Panel on Climate Change (IPPC) concluded that Earth's climate would change dramatically over the next several decades.[15] The report was based on millions of hours of computer simulations on some of the most powerful

---

[14]See David E. Shaw, Martin M. Deneroff, Ron O. Dror, Jeffrey S. Kuskin, Richard H. Larson, John K. Salmon, Cliff Young, Brannon Batson, Kevin J. Bowers, Jack C. Chao, Michael P. Eastwood, Joseph Gagliardo, J. P. Grossman, Richard C. Ho, Douglas J. Lerardi, István Kolossváry, John L. Klepeis, Timothy Layman, Christine Mcleavey, Mark A. Moraes, Rolf Mueller, Edward C. Priest, Yibing Shan, Jochen Spengler, Michael Theobald, Brian Towles, and Stanley C. Wang, 2008, Anton, a special-purpose machine for molecular dynamics simulation, Communications of the ACM 51(7): 91-97.

[15]See IPCC, 2007, Climate Change 2007: Synthesis Report, Contribution of Working Groups I, II and III to the Fourth Assessment Report of the Intergovernmental Panel on Climate Change, eds. Core Writing Team, Rajendra K. Pachauri and Andy Reisinger, Geneva, Switzerland: IPCC. The National Research Council has also recently released three reports noting that strong evidence on climate change underscores the need for actions to reduce emissions and begin adapting to impacts (NRC, 2010, Advancing the Science of Climate Change, Limiting the Magnitude of Climate Change, and Adapting to the Impacts of Climate Change, Washington, D.C.: The National Academies Press, available online at http://www.nap.edu/catalog.php?record_id=12782; NRC, 2010, Limiting the Magnitude of Future Climate Change, Washington, D.C.: The National Academies Press, available online at http://www.nap.edu/catalog.php?record_id=12785; NRC, 2010, Adapting to the Impacts of Climate Change, available online at http://www.nap.edu/catalog.php?record_id=12783.)

supercomputers in the world. The need for computer models in the study of climate change is far from over, however, and the most obvious need is to improve the resolution of models, which previously simulated only one data point every 200 kilometers, whereas physical phenomena like clouds appear on a kilometer scale. To be useful as a predictive tool, climate models need to run at roughly 1,000 times real time, and estimates for a kilometer-scale model therefore require a 20-petaflop (10^{15} floating-point operations per second) machine, which is an order of magnitude faster than the fastest machine available at this writing.[16]

Resolution is only one of the problems, however, and a report based on a DOE workshop suggests that 1 exaflop (10^{18} floating-point operations per second) will be needed within the next decade to meet the needs of the climate research community.[17] Scientists and policy-makers need the increased computational capability to add more features, such as fully resolved clouds, and to capture the potential effects of both natural and human-induced forcing functions on the climate. They also need to understand specific effects of climate change, such as rise in sea levels, changes in ocean circulation, extreme weather events at the local and regional level, and the interaction of carbon, methane, and nitrogen cycles. Climate science is not just about prediction and observation but also about understanding how various regions might need to adapt to changes and how they would be affected by a variety of proposed mitigation strategies. Experimenting with mitigation is expensive, impractical because of time scales, and dangerous—all characteristics that call for improved climate models that can predict favorable and adverse changes in the climate and for improved computing performance to enable such simulations.

Computational Capability and Scientific Progress

The availability of large scientific instruments—such as telescopes, lasers, particle accelerators, and genome sequencers—and of low-cost sensors, cameras, and recording devices has opened up new challenges related to computational analysis of data. Such analysis is useful in a variety of domains. For example, it can be used to observe and understand physical phenomena in space, to monitor air and water quality, to develop a map of the genetic makeup of many species, and to examine alternative

---

[16]See, for example, Olive Heffernan, 2010, Earth science: The climate machine, Nature 463(7284): 1014-1016, which explores the complexity of new Earth models for climate analysis.

[17]DOE, 2009, Scientific Grand Challenges: Challenges in Climate Change Science and the Role of Computing at the Extreme Scale, Workshop Report, Washington D.C., November 6-7, 2008, available online at http://www.er.doe.gov/ascr/ProgramDocuments/Docs/ClimateReport.pdf.

energy sources, such as fusion. Scientific datasets obtained with those devices and simulations are stored in petabyte storage archives in scientific computing centers around the world.

The largest computational problems are often the most visible, but the use of computing devices by individual scientists and engineers is at least as important for the progress of science. Desktop and laptop machines today are an integral part of any scientific laboratory and are used for computations and data-analysis problems that would have required supercomputers only a decade ago. Individual investigators in engineering and science departments around the country use cluster computers based on networks of personal computers. The systems are often shared by a small group of researchers or by larger departments to amortize some of the investment in personnel, infrastructure, and maintenance. And systems that are shared by departments or colleges are typically equivalent in computational capability to the largest supercomputers in the world 5-10 years earlier.

Although no one problem, no matter how large, could have justified the aggregate industry investment that was expended over many years to bring hardware capability up to the required level, we can expect that whole new fields will continue to appear as long as the stream of improvements continues. As another example of the need for more performance in the sciences, the computing centers that run the largest machines have very high use rates, typically over 90 percent, and their sophisticated users study the allocation policies among and within centers to optimize the use of their own allocations. Requests for time on the machines perpetually exceed availability, and anecdotal and statistical evidence suggests that requests are limited by expected availability rather than by characteristics of the science problems to be attacked; when available resources double, so do the requests for computing time.

A slowdown in the growth in computing performance has implications for large swaths of scientific endeavor. The amount of data available and accessible for scientific purposes will only grow, and computational capability needs to keep up if the data are to be used most effectively. Without continued expansion of computing performance commensurate with both the amount of data being generated and the scope and scale of the problems scientists are asked to solve—from climate change to energy independence to disease eradication—it is inevitable that important breakthroughs and opportunities will be missed. Just as in other fields, exponential growth in computing performance has underpinned much scientific innovation. As that growth slows or stops, the opportunities for innovation decrease, and this also has implications for economic competitiveness.

THE IMPORTANCE OF COMPUTING PERFORMANCE FOR DEFENSE AND NATIONAL SECURITY

It is difficult to overstate the importance of IT and computation for defense and national security. The United States has an extremely high-technology military; virtually every aspect depends on IT and computational capability. To expand our capabilities and maintain strategic advantage, operational needs and economic motivators urge a still higher-technology military and national and homeland security apparatus even as many potential adversaries are climbing the same technology curve that we traversed. If, for whatever reason, we do not continue climbing that curve ourselves, the gap between the United States and many of its adversaries will close. This section describes several examples of where continued growth in computing performance is essential for effectiveness. The examples span homeland security, defense, and intelligence and have many obvious nonmilitary applications as well.

Military and Warfighting Needs

There has been no mystery about the efficacy of better technology in weaponry since the longbow first appeared in the hands of English archers or steel swords first sliced through copper shields. World War II drove home the importance of climb rates, shielding, speed, and armament of aircraft and ended with perhaps the most devastating display of unequal armament ever: the nuclear bomb.

The modern U.S. military is based largely on the availability of technologic advantages because it must be capable of maintaining extended campaigns in multiple simultaneous theaters throughout the world, and our armed forces are much smaller than those fielded by several potential adversaries. Technology—such as better communication, satellite links, state-of-the-art weapons platforms, precision air-sea-land launched rockets, and air superiority—acts as a force multiplier that gives the U.S. military a high confidence in its ability to prevail in any conventional fight.

Precision munitions have been a game-changer that enables us to fight wars with far fewer collateral losses than in any recent wars. No longer does the Air Force have to carpet-bomb a section of a city to ensure that the main target of interest is destroyed; instead, it can drop a precision bomb of the required size from a stealthy platform or as a cruise missile from an offshore ship and take out one building on a crowded city street. Sufficiently fast computers provided such capabilities, and faster ones will improve them.

Because high technology has conferred so strong an advantage on the U.S. military for conventional warfare, few adversaries will ever consider engaging us this way. Instead, experiences in Vietnam, Afghanistan, and

Iraq have been unconventional or "asymmetric": instead of large-scale tank engagements, these wars have been conducted on a much more localized basis—between squads or platoons, not brigades or divisions. Since Vietnam, where most of the fighting was in jungles, the venues have been largely urban, in towns where the populace is either neutral or actively hostile. In those settings, the improvised explosive device (IED) has become a most effective weapon for our adversaries.

The military is working on improving methods for detecting and avoiding IEDs, but it is also looking into after-the-fact analysis that could be aided by computer-vision autosurveillance. With remotely piloted aerial vehicles, we can fly many more observation platforms, at lower risk, than ever before. And we can put so many sensors in the air that it is not feasible to analyze all the data that they generate in a timely fashion. At least some part of the analysis must be performed at the source. The better and faster that analysis is, the less real-time human scrutiny is required, and the greater the effectiveness of the devices. A small, efficient military may find itself fighting at tempos that far exceed what was experienced in the past, and this will translate into more sorties per day on an aircraft carrier, faster deployment of ground troops, and quicker reactions to real-time information by all concerned. Coordination among the various U.S. military elements will become much more critical. Using computer systems to manage much of the information associated with these activities could offload the tedious communication and background analytic tasks and move humans' attention to issues for which human judgment is truly required.

Training Simulations

Although often rudimentary, training simulations for the military are ubiquitous and effective. The U.S. government-sponsored first-person-shooter game America's Army is freely downloadable and puts its players through the online equivalent (in weapons, tactics, and first aid) of the training sequence given to a raw recruit. There have been reports that the "training" in the video game America's Army was sufficient to have enabled its players to save lives in the real world.[18] The U.S. Army conducts squad-level joint video-game simulations as a research exercise. Squad tactics, communication, identification of poorly illuminated

---

[18] Reported in Earnest Cavalli, 2008, Man imitates America's army, saves lives, Wired.com, January 18, 2008, available online at http://www.wired.com/gamelife/2008/01/americas-army-t/. The article cites a press release from a game company Web site (The official Army game: America's Army, January 18, 2008, available at http://forum.americasarmy.com/viewtopic.php?t=271086).

targets in houses, and overall movement are stressed and analyzed. In another type of simulation, a single soldier is immersed in a simulation with computer-generated images on all four walls around him. Future training simulations could be made much more realistic, given enough computational capability, by combining accurately portrayed audio (the real sound of real weapons with echos, nulls, and reflections generated by computer) with ever-improving graphics. Humans could be included, perhaps as avatars as in Second Life, who know the languages and customs of the country in which the military is engaged. Limited handheld language-translation devices are being tested in the field; some soldiers like them, and others report difficulty in knowing how and when to use them. Simulations can be run with the same devices so that soldiers can become familiar with the capabilities and limitations and make their use much more efficient. When training simulations become more realistic, they can do what all more accurate simulations do: reduce the need for expensive real-world realizations or increase the range and hazard-level tolerances of operations that would not be possible to train for in the real world.

Autonomous Robotic Vehicles

The Defense Advanced Research Projects Agency (DARPA) has sponsored multiple "Grand Challenge" events in which robotic vehicles compete to traverse a preset course in minimum time. The 2007 competition was in an urban setting in which the vehicles not only were required to stay on the road (and on their own side of the road) but also had to obey all laws and customs that human drivers would. The winning entry, Boss, from the Carnegie Mellon University (CMU) Robotics Institute, had a daunting array of cameras, lidars, and GPS sensors and a massive (for a car) amount of computing horsepower.[19]

CMU's car (and several other competitors) finished the course while correctly identifying many tricky situations on the course, such as the arrival of multiple vehicles at an intersection with stop signs all around. (Correct answer: The driver on the right has the right of way. Unless you got there considerably earlier than that driver, in which case you do. But even if you do have an indisputable right of way, if that driver starts across the intersection, you have the duty to avoid an accident. As it turns out, many humans have trouble with this situation, but the machines largely got it right.)

CMU says that to improve its vehicle, the one thing most desired is

---

[19]See Carnegie Mellon Tartan Racing, available at http://www.tartanracing.org, for more information about the vehicle and the underlying technology.

additional computing horsepower—the more the better. According to DARPA's vision, we appear to be within shouting distance of a robotic military supply truck, one that would no longer expose U.S. military personnel to the threats of IEDs or ambushes. The same technology is also expected to improve civilian transportation and has the potential to reduce collisions on domestic roads and highways.

Domestic Security and Infrastructure

Airport Security Screening

Terrorist groups target civilian populations and infrastructure. The events of 9/11 have sparked many changes in how security is handled, most of which involve computer-based technology. For example, to detect passenger-carried weapons—such as knives, guns, and box cutters—fast x-ray scanners and metal detectors have become ubiquitous in airports throughout the world.

But the x-ray machines are used primarily to produce a two-dimensional image of the contents of carry-on bags; the actual "detector" is the human being sitting in front of the screen. Humans are susceptible to a wide array of malfunctions in that role: they get distracted, they get tired, they get sick, and their effectiveness varies from one person to another. Although it can be argued that there should always be a human in the loop when it is humans one is trying to outsmart, it seems clear that this is an opportunity for increased computational horsepower to augment a human's ability to identify threat patterns in the images.

In the future, one could envision such x-ray image analytic software networking many detectors in an attempt to identify coordinated patterns automatically. Such automation could help to eliminate threats in which a coordinated group of terrorists is attempting to carry on to a plane a set of objects that in isolation are nonthreatening (and will be passed by a human monitor) but in combination can be used in some dangerous way. Such a network might also be used to detect smuggling: one object might seem innocuous, but an entire set carried by multiple people and passed by different screeners might be a pattern of interest to the authorities. And a network might correlate images with weapons found by hand inspection and thus "learn" what various weapons look like when imaged and in the future signal an operator when a similar image appeared.

Surveillance, Smart Cameras, and Video Analytics

A staple of nearly all security schemes is the camera, which typically feeds a real-time low-frame-rate image stream back to a security guard, whose job includes monitoring the outputs of the camera and distinguishing normal situations from those requiring action. As with airport screeners, the security guards find it extremely difficult to maintain the necessary vigilance in monitoring cameras: it is extremely boring, in part because of the low prevalence of the events that they are watching for. But "boring" is what computers do best: they never tire, get distracted, show up for work with a hangover, or fight with a spouse. If a computer system could watch the outputs of cameras 3, 5, and 8 and notify a human if anything interesting happens, security could be greatly enhanced in reliability, scope, and economics.

In the current state of the art, the raw video feed from all cameras is fed directly to monitors with magnetic tape storage or digital sampling to hard drives. With the emergence of inexpensive high-definition cameras, the raw bit rates are quickly climbing well beyond the abilities of networks to transport the video to the monitors economically and beyond the capacity of storage systems to retain the information.

What is needed is for some processing to be performed in the cameras themselves. Suppose that a major retailer needs surveillance on its customer parking lot at night as an antitheft measure. Virtually all the time, the various cameras will see exactly the same scene, down to the last pixel, on every frame, hour after hour. Statistically, the only things that will change from the camera point of view are leaves blowing across the lot, the occasional wild animal, rain, shadows caused by the moon's traversal in the sky, and the general light-dark changes when the sun goes down and comes up the next day. If a camera were smart enough to be able to filter out all the normal, noninteresting events, identifying interesting events would be easier. Although it may be desirable to carry out as much analysis at the camera as possible to reduce the network bandwidth required, the camera may not be constructed in a way that uses much power (for example, it may not have cooling features), and this suggests another way in which power constraints come into play.

Computer technology is only now becoming sophisticated enough at the price and power levels available to a mobile platform to perform some degree of autonomous filtering. Future generations of smart cameras will permit the networking bandwidth freed up by the camera's innate intelligence to be used instead to coordinate observations and decisions made by other cameras and arrive at a global, aggregate situational state of much higher quality than what humans could otherwise have pieced together.

Video search is an important emerging capability in this realm. If you want to find a document on a computer system and cannot remember where it is, you can use the computer system's search feature to help you find it. You might remember part of the file name or perhaps some key words in the document. You might remember the date of creation or the size. All those can be used by the search facility to narrow down the possibilities to the point where you can scan a list and find the one the document that you wanted. Faster computer systems will permit much better automated filtering and searching, and even pictures that have not been predesignated with key search words may still be analyzed for the presence of a person or item of interest.

All the above applies to homeland security as well and could be used for such things as much improved surveillance of ship and aircraft loading areas to prevent the introduction of dangerous items; crowd monitoring at control points; and pattern detection of vehicle movements associated with bombing of facilities.[20]

A related technology is face recognition. It is a very short step from surveilling crowds to asking whether anyone sees a particular face in a crowd and then to asking whether any of a list of "persons of interest" appear in the crowd. Algorithms that are moderately effective in that task already exist. Faster computer systems could potentially improve the accuracy rate by allowing more computation within a given period and increase the speed at which a given frame can be analyzed. As with the overall surveillance problem, networked smart cameras might be able to use correlations to overcome natural-sight impediments.

Infrastructure Defense Against Automated Cyberattack

The Internet now carries a large fraction of all purchases made, so a generalized attack on its infrastructure would cause an immediate loss in sales. Much worse, however, is that many companies and other organizations have placed even their most sensitive documents online, where they are protected by firewalls and virtual private networks but online nonetheless—bits protecting other bits. A coordinated, widespread attack on the U.S. computing and network infrastructure would almost certainly

---

[20]These efforts are much more difficult than it may seem to the uninitiated and, once understood by adversaries, potentially susceptible to countermeasures. For example, England deployed a large set of motorway automated cameras to detect (and deter) speeding; when a camera's radar detected a vehicle exceeding the posted speed limit, the camera snapped a photograph of the offending vehicle and its driver and issued the driver an automated ticket. In the early days of the system's deployment, someone noticed that if the speeding vehicle happened to be changing lanes during the critical period when the radar could have caught it, for some reason the offense would go unpunished. The new lore quickly spread throughout the driving community and led to a rash of inspired lane-changing antics near every radar camera—behavior that was much more dangerous than the speeding would have been. This was reported in Ray Massey, 2006, Drivers can avoid speeding tickets ... by changing lanes, Daily Mail Online, October 15, 2006, available at http://www.dailymail.co.uk/news/article-410539/Drivers-avoid-speeding-tickets--changing-lanes.html.

be set up and initiated via the Internet, and the havoc that it could potentially wreak on businesses and government could be catastrophic. It is not out of the question that such an eventuality could lead to physical war.

The Internet was not designed with security in mind, and this oversight is evident in its architecture and in the difficulty with which security measures can be retrofitted later. We cannot simply dismantle the Internet and start over with something more secure. But as computer-system technology progresses and more performance becomes available, there will be opportunities to look for ways to trade the parallel performance afforded by the technology for improved defensive measures that will discourage hackers, help to identify the people and countries behind cyberattacks, and protect the secrets themselves better.

The global Internet can be a dangerous place. The ubiquitous connectivity that yields the marvelous wonders of search engines, Web sites, browsers, and online purchasing also facilitates identity theft, propagation of worms and viruses, ready platforms for staging denial-of-service attacks, and faceless nearly risk-free opportunities for breaking into the intellectual-property stores and information resources of companies, schools, government institutions, and military organizations. Today, a handful of Web-monitoring groups pool their observations and expertise with a few dozen university computer-science experts and many industrial and government watchdogs to help to spot Internet anomalies, malevolent patterns of behavior, and attacks on the Internet's backbone and name-resolution facilities. As with video surveillance, the battle is ultimately human on human, so it seems unlikely that humans should ever be fully removed from the defensive side of the struggle. However, faster computers can help tremendously, especially if the good guys have much faster computing machinery than the bad guys.

Stateful packet inspection, for example, is a state-of-the-art method for detecting the presence of a set of known virus signatures in traffic on communications networks, which on detection can be shunted into a quarantine area before damage is done. Port-based attacks can be identified before they are launched. The key to those mitigations is that all Internet traffic, harmful or not, must take the form of bits traversing various links of the Internet; computer systems capable of analyzing the contents over any given link are well positioned to eliminate a sizable fraction of threats.

Data Analysis for Intelligence

Vast amounts of unencrypted data not only are not generated in intelligence agencies but are available in the open for strategic data-mining. Continued performance improvements are needed if the agencies are to garner useful intelligence from raw data. There is a continuing need to analyze satellite images for evidence of military and nuclear buildups, evidence of emerging droughts or other natural disasters, evidence of terrorist training camps, and so on. Although it is no secret that the National Security Agency and the National Reconnaissance Office have some of the largest computer complexes in the world, the complexity of the data that they store and process and of the questions that they are asked to address is substantial. Increasing amounts of computational horsepower are needed not only to meet their mission objectives but also to maintain an advantage over adversaries.

Nuclear-Stockpile Stewardship

In the past, the reliability of a nuclear weapon (the probability that it detonates when commanded to do so) and its safety (the probability that it does not detonate otherwise) were established largely with physical testing. Reliability tests detonated sample nuclear weapons from the stockpile, and safety tests subjected sample nuclear weapons to extreme conditions (such as fire and impact) to verify that they did not detonate under such stresses. However, for a variety of policy reasons, the safety and reliability of the nation's nuclear weapons is today established largely with computer simulation, and the data from nonnuclear laboratory experiments are used to validate the computer models.

The simulation of a nuclear weapon is computationally extremely demanding in both computing capability and capacity. The already daunting task is complicated by the need to simulate the effects of aging. A 2003 JASON report[21] concluded that at that time there were gaps in both capability and capacity in fulfilling the mission of stockpile stewardship—ensuring nuclear-weapon safety and reliability.

Historically, the increase in single-processor performance played a large role in providing increased computing capability and capacity to meet the increasing demands of stockpile stewardship. In addition, parallelism has been applied to the problem, so the rate of increase in performance of the large machines devoted to the task has been greater than called for by Moore's law because the number of processors was increased at the same time that single-processor performance was increasing. The largest of the machines today have over 200,000 processors and LINPACK benchmark performance of more than 1,000 Tflops.[22]

---

[21]Roy Schwitters, 2003, Requirements for ASCI, JSR-03-330, McLean, Va.: The MITRE Corporation.

[22]For a list of the 500 most powerful known computer systems in the world, see "Top 500," available online at http://www.absoluteastronomy.com/topics/TOP500.

The end of single-processor performance scaling makes it difficult for those "capability" machines to continue scaling at historical rates and so makes it difficult to meet the projected increases in demands of nuclear-weapon simulation. The end of single-processor scaling has also made the energy and power demands of future capability systems problematic, as described in the recent DARPA ExaScale computing study.[23] Furthermore, the historical increases in demand in the consumer market for computing hardware and software have driven down costs and increased software capabilities for military and science applications. If the consumer market suffers, the demands of science and military applications are not likely to be met.

THE IMPORTANCE OF COMPUTING PERFORMANCE FOR CONSUMER NEEDS AND APPLICATIONS

The previous two sections offered examples of where growth in computing performance has been essential for science, defense, and national security. The growth has also been a driver for individuals using consumer-oriented systems and applications. Two recent industry trends have substantially affected end-user computational needs: the increasing ubiquity of digital data and growth in the population of end users who are not technically savvy. Sustained growth in computing performance serves not only broad public-policy objectives, such as a strong defense and scientific leadership, but also the current and emerging needs of individual users.

The growth in computing performance over the last 4 decades—impressive though it has been—has been dwarfed over the last decade or so by the growth in digital data.[24] The amount of digital data is growing more rapidly than ever before. The volumes of data now available outstrip our ability to comprehend it, much less take maximum advantage

---

[23]Peter Kogge, Keren Bergman, Shekhar Borkar, Dan Campbell, William Carlson, William Dally, Monty Denneau, Paul Franzon, William Harrod, Kerry Hill, Jon Hiller, Sherman Karp, Stephen Keckler, Dean Klein, Robert Lucas, Mark Richards, Al Scarpelli, Steven Scott, Allan Snavely, Thomas Sterling, R. Stanley Williams, and Katherine Yelick, 2008, ExaScale Computing Study: Technology Challenges in Achieving Exascale Systems, Washington, D.C.: DARPA. Available online at http://www.er.doe.gov/ascr/Research/CS/DARPA%20 exascale%20-%20hardware%20(2008).pdf.

[24]A February 2010 report observed that "quantifying the amount of information that exists in the world is hard. What is clear is that there is an awful lot of it, and it is growing at a terrific rate (a compound annual 60%) that is speeding up all the time. The flood of data from sensors, computers, research labs, cameras, phones and the like surpassed the capacity of storage technologies in 2007" (Data, data, everywhere: A special report on managing information, The Economist, February 25, 2010, available online at http://www.economist.com/displaystory.cfm?story_id=15557443).

of it. According to the *How Much Information* project at the University of California, Berkeley,[25] print, film, magnetic, and optical storage media produced about 5 exabytes (EB) of new information in 2003. Furthermore, the information explosion is accelerating. Market research firm IDC estimates that in 2006 161 EB of digital content was created and that that figure will rise to 988 EB by 2010. To handle so much information, people will need systems that can help them to understand the available data. We need computers to see data the way we do, identify what is useful to us, and assemble it for our review or even process it on our behalf. This growing end-user need is the primary force behind the radical and continuing transformation of the Web as it shifts its focus from data presentation to end-users to automatic data-processing on behalf of end-users.[26] The data avalanche and the consequent transformation of the Web's functionality require increasing sophistication in data-processing and hence additional computational capability to be able to reason automatically in real time so that we can understand and interpret structured and unstructured collections of information via, for example, sets of dynamically learned inference rules.

A computer's ability to perform a huge number of computations per second has enabled many applications that have an important role in our daily lives.[27] An important subset of applications continues to push the frontiers of very high computational needs. Examples of such applications are these:

- Digital content creation—allows people to express creative skills and be entertained through various modern forms of electronic arts, such as animated films, digital photography, and video games.
- Search and mining—enhances a person's ability to search and recall objects, events, and patterns well beyond the natural limits of human memory by using modern search engines and the ever-growing archive of globally shared digital content.

---

[25] See Peter Lyman and Hal R. Varian, 2003, How much information?, available online at http://www2.sims.berkeley.edu/research/projects/how-much-info-2003/index.htm, last accessed November 2, 2010.

[26] See, for example, Tim Berners-Lee's 2007 testimony to the U.S. Congress on the future of the World Wide Web, The digital future of the United States. Part I: The future of the World Wide Web," Hearings before the Subcommittee on Telecommunications and the Internet of the Committee on Energy and Commerce, 110th Congress, available at http://dig.csail.mit.edu/2007/03/01-ushouse-future-of-the-web.html, last accessed November 2, 2010.

[27] Of course, a computer system's aggregate performance may be limited by many things: the nature of the workload itself, the CPU's design, the memory subsystem, input/output device speeds and sizes, the operating system, and myriad other system aspects. Those and other aspects of performance are discussed in Chapter 2.

- Real-time decision-making—enables growing use of computational assistance for various complex problem-solving tasks, such as speech transcription and language translation.
- Collaboration technology—offers a more immersive and interactive 3D environment for real-time collaboration and telepresence.
- Machine-learning algorithms—filter e-mail spam, supply reliable telephone-answering services, and make book and music recommendations.

Computers have become so pervasive that a vast majority of end-users are not computer aficionados or system experts; rather, they are experts in some other field or disciplines, such as science, art, education, or entertainment. The shift has challenged the efficiency of human-computer interfaces. There has always been an inherent gap between a user's conceptual model of a problem and a computer's model of the problem. However, given the change in demographics of computer users, the need to bridge the gap is now more acute than ever before. The increased complexity of common end-user tasks ("find a picture like this" rather than "add these two numbers") and the growing need to be able to offer an effective interface to a non-computer-expert user at a higher level of object semantics (for example, presenting not a Fourier transform data dump of a flower image but a synthesized realistic visual of a flower) have together increased the computational capability needed to provide real-time responses to user actions.

Bridging the gap would be well served by computers that can deal with natural user inputs, such as speech and gestures, and output content in a visually rich form close to that of the physical world around us. A typical everyday problem requires multiple iterations of *execute and evaluate* between the user and the computer system. Each such iteration normally narrows the original modeling gap, and this in turn requires additional computational capability. The larger the original gap, the more computation is needed to bridge it. For example, some technology-savvy users working on an image-editing problem may iterate by editing a low-level machine representation of an image, whereas a more typical end-user may interact only at the level of a photo-real output of the image with virtual brushes and paints.

Thanks to sustained growth in computing performance over the years, more effective computer-use models and visually rich human-computer interfaces are introducing new potential ways to bridge the gap. An alternative to involving the end-user in each iteration is to depend on a computer's ability to refine model instances by itself and to nest multiple iterations of such an *analytics* loop for each iteration of a *visual computing* loop involving an end-user. Such nesting allows a reduction in the number of interactions between a user and his or her computer and therefore an increase in the system's efficiency or response. However, it also creates the need to sustain continued growth in computational performance so that a wider variety of more complex tasks can be simulated and solved in real time for the growing majority of end-users. Real-time physical and behavioral simulation of even a simple daily-life object or events (such as water flow, the trajectory of a ball in a game, and summarizing of a text) is a surprisingly computationally expensive task, and requires multiple iterations or solutions of a large number of subproblems derived from decomposition of the original problem.

Computationally intensive consumer applications include such phenomena as virtual world simulations and immersive social-networking, video karaoke (and other sorts of real-time video interactions), remote education and training that require simulation, and telemedicine (including interventional medical imaging).[28]

THE IMPORTANCE OF COMPUTING PERFORMANCE FOR ENTERPRISE PRODUCTIVITY

Advances in computing technology in the form of more convenient communication and sharing of information have favorably affected the productivity of enterprises. Improved communication and sharing have been hallmarks of computing from the earliest days of time-sharing in corporate or academic environments to today's increasingly mobile, smart phone-addicted labor force. Younger employees in many companies today can hardly recall business processes that did not make use of e-mail, chat and text messaging, group calendars, internal Web resources, blogs, Wiki toolkits, audio and video conferencing, and automated management of workflow. At the same time, huge improvements in magnetic storage technology, particularly for disk drives, have made it affordable to keep every item of an organization's information accessible on line. Individual worker productivity is not the only aspect of an enterprise that has been affected by continued growth in computing performance. The ability of virtually every sort of enterprise to use computation to understand data related to its core lines of business—sometimes referred to as analytics— has improved dramatically as computer performance has increased over the years. In addition, massive amounts of data and computational capability accessible on the Internet have increased the demand for Web services, or "software as a service," in a variety of sectors. Analytics and

---

[28]For more on emerging applications and their need for computational capability, see Justin Rattner, 2009, The dawn of terascale computing, IEEE Solid-State Circuits Magazine 1(1): 83-89.

the implications of Web services for computing performance needs are discussed below.

Analytics

Increases in computing capability and efficiency have made it feasible to perform deep analysis of numerous kinds of business data—not just off line but increasingly in real time—to obtain better input into business decisions.[29] Efficient computerized interactions between organizations have created more efficient end-to-end manufacturing processes through the use of supply-chain management systems that optimize inventories, expedite product delivery, and reduce exposure to varying market conditions.

In the past, real-time business performance needs were dictated mostly by transaction rates. Analytics (which can be thought of as computationally enhanced decision-making) were mostly off line. The computational cost of actionable data-mining was too high to be of any value under real-time use constraints. However, the growth in computing performance has now made real-time analytics affordable for a larger class of enterprise users.

One example is medical-imaging analytics. Over the last 2 decades, unprecedented growth has taken place in the amount and complexity of digital medical-image data collected on patients in standard medical practice. The clinical necessity to diagnose diseases accurately and develop treatment strategies in a minimally invasive manner has mandated the development of new image-acquisition methods, high-resolution acquisition hardware, and novel imaging modalities. Those requirements have placed substantial computational burdens on the ability to use the image information synergistically. With the increase in the quality and utility of medical-image data, clinicians are under increasing pressure to generate more accurate diagnoses or therapy plans. To meet the needs of the clinician, the imaging-research community must provide real-time (or near real-time) high-volume visualization and analyses of the image data to optimize the clinical experience. Today, nearly all use of computation in medical imaging is limited to "diagnostic imaging." However, with sufficient computational capability, it is likely that real-time medical interventions could become possible. The shift from diagnostic imaging to interventional imaging can usher in a new era in medical imaging. Real-time

---

[29]IBM's Smart Analytics System, for example, is developing solutions aimed at retail, insurance, banking, health care, and telecommunication. For more information see the IBM Smart Analytics System website, available online at http://www-01.ibm.com/software/data/infosphere/smart-analytics-system/.

medical analytics can guide medical professionals, such as surgeons, in their tasks. For example, surface extractions from volumetric data coupled with simulations of various what-if scenarios accomplished in real time offer clear advantages over basic preoperative planning scenarios.

Web Services

In the last 15 years, the Internet and the Web have had a transformational effect on people's lives. That effect has been enabled by two concurrent and interdependent phenomena: the rapid expansion of Internet connectivity, particularly high-speed Internet connections, and the emergence of several extraordinarily useful Internet-based services. Web search and free Web-based e-mail were among the first such services to explode in popularity, and their emergence and continuous improvements have been made possible by dramatic advances in computing performance, storage, and networking technologies. Well beyond text, Web-server data now include videos, photos, and various other kinds of media. Users—individuals and businesses—increasingly need information systems to *see* data the way they do, identify what is useful, and assemble it for them. The ability to have computers understand the data and help us to use it in various enterprise endeavors could have enormous benefits. As a result, the Web is shifting its focus from data presentation to end-users to automatic data-processing on behalf of end-users. Finding preferred travel routes while taking real-time traffic feeds into account and rapid growth in program trading are some of the examples of real-time decision-making.

Consider Web search as an example. A Web search service's fundamental task is to take a user's query, traverse data structures that are effectively proportional in size to the total amount of information available on line, and decide how to select from among possibly millions of candidate results the handful that would be most likely to match the user's expectation. The task needs to be accomplished in a few hundred milliseconds in a system that can sustain a throughput of several thousand requests per second. This and many other Web services are offered free and rely on on-line advertisement revenues, which, depending on the service, may bring only a few dollars for every thousand user page views. The computing system that can meet those performance requirements needs to be not only extremely powerful but also extremely cost-efficient so that the business model behind the Internet service remains viable.

The appetite of Internet services for additional computing performance doesn't appear to have a foreseeable limit. A Web search can be used to illustrate that, although a similar rationale could be applied to other types of services. Search-computing demands fundamentally grow in three dimensions: data-repository increases, search-query increases, and service-quality improvements. The amount of information currently indexed by search engines, although massive, is still generally considered a fraction of all on-line content even while the Web itself keeps expanding. Moreover, there are still several non-Web data sources that have yet to be added to the typical Web-search repositories (such as printed media). Universal search,[30] for example, is one way in which search-computing demands can dramatically increase as all search queries are simultaneously sent to diverse data sources. As more users go online or become more continuously connected to the Internet through better wireless links, traffic to useful services would undergo further substantial increases.

In addition to the amount of data and types of queries, increases in the quality of the search product invariably cause more work to be performed on behalf of each query. For example, better results for a user's query will often be satisfied by searching also for some common synonyms or plurals of the original query terms entered. To achieve the better results, one will need to perform multiple repository lookups for the combinations of variations and pick the best results among them, a process that can easily increase the computing demands for each query by substantial factors.

In some cases, substantial service-quality improvements will demand improvements in computing performance along multiple dimensions simultaneously. For example, the Web would be much more useful if there were no language barriers; all information should be available in every existing language, and this might be achievable through machine-translation technology at a substantial processing cost. The cost would come both from the translation step itself, because accurate translations require very large models or learning over large corpora, and from the increased amount of information that then becomes available for users of every language. For example, a user search in Italian would traverse not only Italian-language documents but potentially documents in every language available to the translation system. The benefits to society at large from overcoming language barriers would arguably rival any other single technologic achievement in human history, especially if they extended to speech-to-speech real-time systems.

The prospect of mobile computing systems—such as cell phones, vehicle computers, and media players—that are increasingly powerful, ubiquitous, and interconnected adds another set of opportunities for bet-

---

[30]See Google's announcement: Google begins move to universal search: Google introduces new search features and unveils new homepage design," Press Release, Google.com, May 16, 2007, available online at http://www.google.com/intl/en/press/pressrel/universal-search_20070516.html.

ter computing services that go beyond simply accessing the Web on more devices. Such devices could act as useful sensors and provide a rich set of data about their environment that could be useful once aggregated for real-time disaster response, traffic-congestion relief, and as-yet-unimagined applications. An early example of the potential use of such systems is illustrated in a recent experiment conducted by the University of California, Berkeley, and Nokia in which cell phones equipped with GPS units were used to provide data for a highway-conditions service.[31]

More generally, the unabated growth in digital data, although still a challenge for managing and sifting, has now reached a data volume large enough in many cases to have radical computing implications.[32] Such huge amounts of data will be especially useful for a class of problems that have so far defied analytic formulation and been reliant on a statistical data-driven approach. In the past, because of insufficiently large datasets, the problems have had to rely on various, sometimes questionable heuristics. Now, the digital-data volume for many of the problems has reached a level sufficient to revert to statistical approaches. Using statistical approaches for this class of problems presents an unprecedented opportunity in the history of computing: the intersection of massive data with massive computational capability.

In addition to the possibility of solving problems that have heretofore been intractable, the massive amounts of data that are increasingly available for analysis by small and large businesses offer the opportunity to develop new products and services based on that analysis. Services can be envisioned that automate the analysis itself so that the businesses do not have to climb this learning curve. The machine-learning community has many ideas for quasi-intelligent automated agents that can roam the Web and assemble a much more thorough status of any topic at a much deeper level than a human has time or patience to acquire. Automated inferences can be drawn that show connections that have heretofore been unearthed only by very talented and experienced humans.

On top of the massive amounts of data being created daily and all that portends for computational needs, the combination of three elements has the potential to deliver a massive increase in real-time computational resources targeted toward end-user devices constrained by cost and power:

---

[31]See the University of California, Berkeley, press release about this experiment (Sarah Yang, 2008, Joint Nokia research project captures traffic data using GPS-enabled cell phones, Press Release, UC Berkeley News, February 8, 2008, available online at http://berkeley.edu/news/media/releases/2008/02/08_gps.shtml).

[32]Wired.com ran a piece in 2008 declaring "the end of science": The Petabyte Age: Because more isn't just more—more is different," Wired.com, June 23, 2008, available online at http://www.wired.com/wired/issue/16-07.

- Clouds of servers.
- Vastly larger numbers of end-user devices, consoles, and various form-factor computing platforms.
- The ubiquitous connectivity of computing equipment over a service-oriented infrastructure backbone.

The primary technical challenge to take advantage of those resources lies in software. Specifically, innovation is needed to enable the discovery of the computing needs of various functional components of a specific service offering. Such discovery is best done adaptively and under the real-time constraints of available computing bandwidth at the client-server ends, network bandwidth, and latency. On-line games, such as *Second Life,* and virtual world simulations, such as *Google Earth,* are examples of such a service. The services involve judicious decomposition of computing needs over public client-server networks to produce an interactive, visually rich end-user experience. The realization of such a vision of *connected computing* will require not only increased computing performance but standardization of network software layers. Standardization should make it easy to build and share unstructured data and application programming interfaces (APIs) and enable ad hoc and innovative combinations of various service offerings.

In summary, computing in a typical end-user's life is undergoing a momentous transformation from being useful yet nonessential software and products to being the foundation for around-the-clock relied-on vital services delivered by tomorrow's enterprises.

2

What Is Computer Performance?

Fast, inexpensive computers are now essential to numerous human endeavors. But less well understood is the need not just for fast computers but also for ever-faster and higher-performing computers at the same or better costs. Exponential growth of the type and scale that have fueled the entire information technology industry is ending.[1] In addition, a growing performance gap between processor performance and memory bandwidth, thermal-power challenges and increasingly expensive energy use, threats to the historical rate of increase in transistor density, and a broad new class of computing applications pose a wide-ranging new set of challenges to the computer industry. Meanwhile, societal expectations for increased technology performance continue apace and show no signs of slowing, and this underscores the need for ways to sustain exponentially increasing performance in multiple dimensions. The essential engine that has met this need for the last 40 years is now in considerable danger, and this has serious implications for our economy, our military, our research institutions, and our way of life.

Five decades of exponential growth in processor performance led to

---

[1] It can be difficult even for seasoned veterans to understand the effects of exponential growth of the sort seen in the computer industry. On one level, industry experts, and even consumers, display an implicit understanding in terms of their approach to application and system development and their expectations of and demands for computing technologies. On another level, that implicit understanding makes it easy to overlook how extraordinary the exponential improvements in performance of the sort seen in the information technology industry actually are.

the rise and dominance of the general-purpose personal computer. The success of the general-purpose microcomputer, which has been due primarily to economies of scale, has had a devastating effect on the development of alternative computer and programming models. The effect can be seen in high-end machines like supercomputers and in low-end consumer devices, such as media processors. Even though alternative architectures and approaches might have been technically superior for the task they were built for, they could not easily compete in the marketplace and were readily overtaken by the ever-improving general-purpose processors available at a relatively low cost. Hence, the personal computer has been dubbed "the killer micro."

Over the years, we have seen a series of revolutions in computer architecture, starting with the main-frame, the minicomputer, and the work station and leading to the personal computer. Today, we are on the verge of a new generation of smart phones, which perform many of the applications that we run on personal computers and take advantage of network-accessible computing platforms (cloud computing) when needed. With each iteration, the machines have been lower in cost per performance and capability, and this has broadened the user base. The economies of scale have meant that as the per-unit cost of the machine has continued to decrease, the size of the computer industry has kept growing because more people and companies have bought more computers. Perhaps even more important, general-purpose single processors—which all these generations of architectures have taken advantage of—can be programmed by using the same simple, sequential programming abstraction at root. As a result, software investment on this model has accumulated over the years and has led to the de facto standardization of one instruction set, the Intel x86 architecture, and to the dominance of one desktop operating system, Microsoft Windows.

The committee believes that the slowing in the exponential growth in computing performance, while posing great risk, may also create a tremendous opportunity for innovation in diverse hardware and software infrastructures that excel as measured by other characteristics, such as low power consumption and delivery of throughput cycles. In addition, the use of the computer has becomes so pervasive that it is now economical to have many more varieties of computers. Thus, there are opportunities for major changes in system architectures, such as those exemplified by the emergence of powerful distributed, embedded devices, that together will create a truly ubiquitous and invisible computer fabric. Investment in whole-system research is needed to lay the foundation of the computing environment for the next generation. See Figure 2.1 for a graph showing flattening curves of performance, power, and frequency.

Traditionally, computer architects have focused on the goal of creating

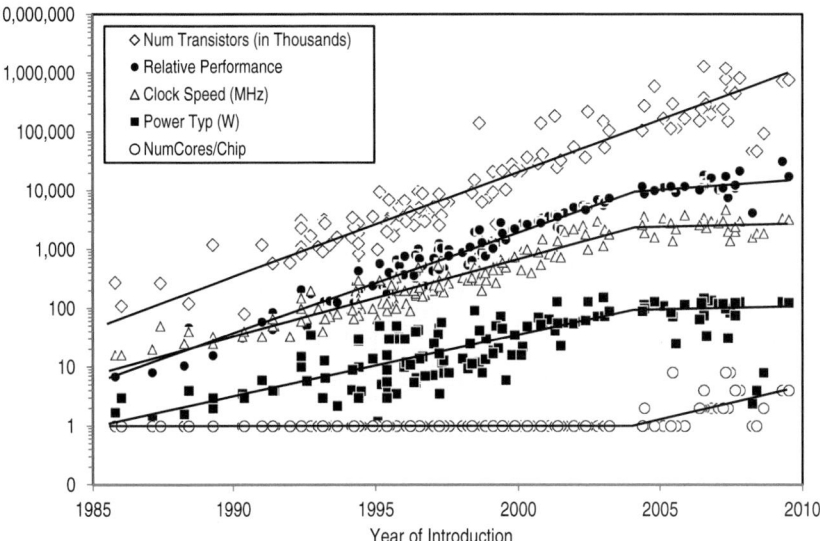

FIGURE 2.1 Transistors, frequency, power, performance, and cores over time (1985-2010). The vertical scale is logarithmic. Data curated by Mark Horowitz with input from Kunle Olukotun, Lance Hammond, Herb Sutter, Burton Smith, Chris Batten, and Krste Asanoviç.

computers that perform single tasks as fast as possible. That goal is still important. Because the uniprocessor model we have today is extremely powerful, many performance-demanding applications can be mapped to run on networks of processors by dividing the work up at a very coarse granularity. Therefore, we now have great building blocks that enable us to create a variety of high-performance systems that can be programmed with high-level abstractions. There is a serious need for research and education in the creation and use of high-level abstractions for parallel systems.

However, single-task performance is no longer the only metric of interest. The market for computers is so large that there is plenty of economic incentive to create more specialized and hence more cost-effective machines. Diversity is already evident. The current trend of moving computation into what is now called the cloud has created great demands for high-throughput systems. For those systems, making each transaction run as fast as possible is not the best thing to do. It is better, for example, to have a larger number of lower-speed processors to optimize the throughput rate and minimize power consumption. It is similarly important to conserve power for hand-held devices. Thus, power consumption is a

> **BOX 2.1**
> **Embedded Computing Performance**
>
> The design of desktop systems often places considerable emphasis on general CPU performance in running desktop workloads. Particular attention is paid to the graphics system, which directly determines which consumer games will run and how well. Mobile platforms, such as laptops and notebooks, attempt to provide enough computing horsepower to run modern operating systems well—subject to the energy and thermal constraints inherent in mobile, battery-operated devices—but tend not to be used for serious gaming, so high-end graphics solutions would not be appropriate. Servers run a different kind of workload from either desktops or mobile platforms, are subject to substantially different economic constraints in their design, and need no graphics support at all. Desktops and mobile platforms tend to value legacy compatibility (for example, that existing operating systems and software applications will continue to run on new hardware), and this compatibility requirement affects the design of the systems, their economics, and their use patterns.
>
> Although desktops, mobile, and server computer systems exhibit important differences from one another, it is natural to group them when comparing them with embedded systems. It is difficult to define embedded systems accurately because their space of applicability is huge—orders of magnitude larger than the general-purpose computing systems of desktops, laptops, and servers. Embedded computer systems can be found everywhere: a car's radio, engine controller, transmission controller, airbag deployment, antilock brakes, and dozens of other places. They are in the refrigerator, the washer and dryer, the furnace controller, the MP3 player, the television set, the alarm clock, the treadmill and stationary bike, the Christmas lights, the DVD player, and the power tools in the garage. They might even be found in ski boots, tennis shoes, and greeting cards. They control the elevators and heating and cooling systems at the office, the video surveillance system in the parking lot, and the lighting, fire protection, and security systems.
>
> Every computer system has economic constraints. But the various systems tend to fall into characteristic financial ranges. Desktop systems once (in 1983) cost $3,000 and now cost from a few hundred dollars to around $1,000. Mobile systems cost more at the high end, perhaps $2,500, down to a few hundred dollars at the low end. Servers vary from a few thousand dollars up to hundreds of thousands for a moderate Web server, a few million dollars for a small corporate

key performance metric for both high-end servers and consumer handheld devices. See Box 2.1 for a discussion of embedded computing performance as distinct from more traditional desktop systems. In general, power considerations are likely to lead to a large variety of specialized processors.

The rest of this chapter provides the committee's views on matters related to computer performance today. These views are summarized in

server farm, and 1 or 2 orders of magnitude more than that for the huge server farms fielded by large companies, such as eBay, Yahoo!, and Google.

Embedded systems tend to be inexpensive. The engine controller under the hood of a car cost the car manufacturer about $3-5. The chips in a cell phone were also in that range. The chip in a tennis shoe or greeting card is about 1/10 that cost. The embedded system that runs such safety-critical systems as elevators will cost thousands of dollars, but that cost is related more to the system packaging, design, and testing than to the silicon that it uses.

One of the hallmarks of embedded systems versus general-purpose computers is that, unlike desktops and servers, embedded performance is not an open-ended boon. Within their cost and power budgets, desktops, laptops, and server systems value as much performance as possible—the more the better. Embedded systems are not generally like that. The embedded chip in a cell phone has a set of tasks to perform, such as monitoring the phone's buttons, placing various messages and images on the display, controlling the phone's energy budget and configuration, and setting up and receiving calls. To accomplish those tasks, the embedded computer system (comprising a central processor, its memory, and I/O facilities) must be capable of a some overall performance level. The difference from general-purpose computers is that once that level is reached in the system design, driving it higher is not beneficial; in fact, it is detrimental to the system. Embedded computer systems that are faster than necessary to meet requirements use more energy, dissipate more heat, have lower reliability, and cost more—all for no gain.

Does that mean that embedded processors are now fast enough and have no need to go faster? Are they exempt from the emphasis in this report on "sustaining growth in computing performance"? No. If embedded processor systems were to become faster and all else were held equal, embedded-system designers would find ways of using the additional capability, and delivering new functionalities would come to be expected on those devices. For example, many embedded systems, such as the GPS or audio system in a car, tend to interface directly with human beings. Voice and speech recognition capability greatly enhance that experience, but current systems are not very good at the noise suppression, beam-forming, and speech-processing that are required to make this a seamless, enjoyable experience, although progress is being made. Faster computer systems would help to solve that problem. Embedded systems have benefited tremendously from riding an improvement curve equivalent to that of the general-purpose systems and will continue to do so in the future.

the bullet points that follow this paragraph. Readers who accept the committee's views may choose to skip the supporting arguments and move on to the next chapter.

- Increasing computer performance enhances human productivity.
- One measure of single-processor performance is the product of operating frequency, instruction count, and instructions per cycle.

- Performance comes directly from faster devices and indirectly from using more devices in parallel.
- Parallelism can be helpfully divided into instruction-level parallelism, data-level parallelism, and thread-level parallelism.
- Instruction-level parallelism has been extensively mined, but there is now broad interest in data-level parallelism (for example, due to graphics processing units) and thread-level parallelism (for example, due to chip multiprocessors).
- Computer-system performance requires attention beyond processors to memories (such as,dynamic random-access memory), storage (for example, disks), and networking.
- Some computer systems seek to improve responsiveness (for example, timely feedback to a user's request), and others seek to improve throughput (for example, handling many requests quickly).
- Computers today are implemented with integrated circuits (chips) that incorporate numerous devices (transistors) whose population (measured as transistors per chaip) has been doubling every 1.5-2 years (Moore's law).
- Assessing the performance delivered to a user is difficult and depends on the user's specific applications.
- Large parts of the potential performance gain due to device innovations have been usefully applied to productivity gains (for example, via instruction-set compatibility and layers of software).
- Improvements in computer performance and cost have enabled creative product innovations that generated computer sales that, in turn, enabled a virtuous cycle of computer and product innovations.

WHY PERFORMANCE MATTERS

Humans design machinery to solve problems. Measuring how well machines perform their tasks is of vital importance for improving them, conceiving better machines, and deploying them for economic benefit. Such measurements often speak of a machine's *performance*, and many aspects of a machine's operations can be characterized as performance. For example, one aspect of an automobile's performance is the time it takes to accelerate from 0 to 60 mph; another is its average fuel economy. Braking ability, traction in bad weather conditions, and the capacity to tow trailers are other measures of the car's performance.

Computer systems are machines designed to perform information processing and computation. Their performance is typically measured by how much information processing they can accomplish per unit time, but there are various perspectives on what type of information processing to consider when measuring performance and on the right time scale for such measurements. Those perspectives reflect the broad array of uses and the diversity of end users of modern computer systems. In general, the systems are deployed and valued on the basis of their ability to improve *productivity*. For some users, such as scientists and information technology specialists, the improvements can be measured in quantitative terms. For others, such as office workers and casual home users, the performance and resulting productivity gains are more qualitative. Thus, no single measure of performance or productivity adequately characterizes computer systems for all their possible uses.[2]

On a more technical level, modern computer systems deploy and coordinate a vast array of hardware and software technologies to produce the results that end users observe. Although the raw computational capabilities of the central processing unit (CPU) core tend to get the most attention, the reality is that performance comes from a complex balance among many cooperating subsystems. In fact, the underlying performance bottlenecks of some of today's most commonly used large-scale applications, such as Web searching, are dominated by the characteristics of memory devices, disk drives, and network connections rather than by the CPU cores involved in the processing. Similarly, the interactive responsiveness perceived by end users of personal computers and hand-held devices is typically defined more by the characteristics of the operating system, the graphical user interface (GUI), and the storage components than by the CPU core. Moreover, today's ubiquitous networking among computing devices seems to be setting the stage for a future in which the computing experience is defined at least as much by the coordinated interaction of multiple computers as it is by the performance of any node in the network.

Nevertheless, to understand and reason about performance at a high level, it is important to understand the fundamental lower-level contributors to performance. CPU performance is the driver that forces the many other system components and features that contribute to overall performance to keep up and avoid becoming bottlenecks

PERFORMANCE AS MEASURED BY RAW COMPUTATION

The classic formulation for raw computation in a single CPU core identifies *operating frequency, instruction count,* and *instructions per cycle*

---

[2]Consider the fact that the term "computer system" today encompasses everything from small handheld devices to Netbooks to corporate data centers to massive server farms that offer cloud computing to the masses.

(IPC) as the fundamental low-level components of performance.[3] Each has been the focus of a considerable amount of research and discovery in the last 20 years. Although detailed technical descriptions of them are beyond the intended scope of this report, the brief descriptions below will provide context for the discussions that follow.

- *Operating frequency* defines the basic clock rate at which the CPU core runs. Modern high-end processors run at several billion cycles per second. Operating frequency is a function of the low-level transistor characteristics in the chip, the length and physical characteristics of the internal chip wiring, the voltage that is applied to the chip, and the degree of pipelining used in the microarchitecture of the machine. The last 15 years have seen dramatic increases in the operating frequency of CPU cores. As an unfortunate side effect of that growth, the maximum operating frequency has often been used as a proxy for performance by much of the popular press and industry marketing campaigns. That can be misleading because there are many other important low-level and system-level measures to consider in reasoning about performance.
- *Instruction count* is the number of native instructions—instructions written for that specific CPU—that must be executed by the CPU to achieve correct results with a given computer program. Users typically write programs in high-level programming languages—such as Java, C, C++, and C#—and then use a *compiler* to translate the high-level program to native machine instructions. Machine instructions are specific to the *instruction set architecture* (ISA) that a given computer architecture or architecture family implements. For a given high-level program, the machine instruction count varies when it executes on different computer systems because of differences in the underlying ISA, in the microarchitecture that implements the ISA, and in the tools used to compile the program. Although this section of the report focuses mostly on the low-level raw performance measures, the role of the compiler and other modern software system technologies are also necessary to understand performance fully.
- *Instructions per cycle* refers to the average number of instructions that a particular CPU core can execute and complete in each cycle. IPC is a strong function of the underlying microarchitecture, or machine organization, of the CPU core. Many modern CPU

---

[3]John L. Hennessy and David A. Patterson, 2006, Computer Architecture: A Quantitative Approach, fourth edition, San Francisco, Cal.: Morgan Kauffman.

cores use advanced techniques—such as multiple instruction dispatch, out-of-order execution, branch prediction, and speculative execution—to increase the average IPC.[4] Those techniques all seek to execute multiple instructions in a single cycle by using additional resources to reduce the total number of cycles needed to execute the program. Some performance assessments focus on the peak capabilities of the machines; for example, the peak performance of the IBM Power 7 is six instructions per cycle, and that of the Intel Pentium, four. In reality, those and other sophisticated CPU cores actually sustain an average of slightly more than one instruction per cycle when executing many programs. The difference between theoretical peak performance and actual sustained performance is an important aspect of overall computer-system performance.

The program itself provides different forms of *parallelism* that different machine organizations can exploit to achieve performance. The first type, *instruction-level parallelism*, describes the amount of nondependent instructions[5] available for parallel execution at any given point in the program. The program's instruction-level parallelism in part determines the IPC component of raw performance mentioned above. (IPC can be viewed as describing the degree to which a particular machine organization can harvest the available instruction-level performance.) The second type of parallelism is *data-level parallelism,* which has to do with how data elements are distributed among computational units for similar types of processing. Data-level parallelism can be exploited through architectural and microarchitectural techniques that direct low-level instructions to operate on multiple pieces of data at the same time. This type of processing is often referred to as single-instruction-multiple-data. The third type is *thread-level parallelism* and has to do with the degree to which a program can be partitioned into multiple sequences of instructions with the intent of executing them concurrently and cooperatively on multiple processors. To exploit program parallelism, the compiler or run-time system must map it to appropriate parallel hardware.

Throughout the history of modern computer architecture, there have been many attempts to build machines that exploit the various forms of

---

[4]Providing the details of these microarchitecture techniques is beyond the scope of this publication. See Hennessey & Patterson for more information on these and related techniques.

[5]An instruction X does not depend on instruction Y if X can be performed without using results from Y. The instruction a = b + c depends on previous instructions that produce the results b and c and thus cannot be executed until those previous instructions have completed.

parallelism. In recent years, owing largely to the emergence of more generalized and programmable forms of graphics processing units, the interest in building machines that exploit data-level parallelism has grown enormously. The specialized machines do not offer compatibility with existing programs, but they do offer the promise of much more performance when presented with code that properly exposes the available data-level parallelism. Similarly, because of the emergence of chip multiprocessors, there is considerable renewed interest in understanding how to exploit thread-level parallelism on these machines more fully. However, the techniques also highlight the importance of the full suite of hardware components in modern computer systems, the communication that must occur among them, and the software technologies that help to automate application development in order to take advantage of parallelism opportunities provided by the hardware.

COMPUTATION AND COMMUNICATION'S EFFECTS ON PERFORMANCE

The raw computational capability of CPU cores is an important component of system-level performance, but it is by no means the only one. To complete any useful tasks, a CPU core must communicate with memory, a broad array of input/output devices, other CPU cores, and in many cases other computer systems. The overhead and latency of that communication in effect delays computational progress as the CPU waits for data to arrive and for system-level interlocks to clear. Such delays tend to reduce peak computational rates to effective computational rates substantially. To understand effective performance, it is important to understand the characteristics of the various forms of communication used in modern computer systems.

In general, CPU cores perform best when all their operands (the inputs to the instructions) are stored in the architected registers that are internal to the core. However, in most architectures, there tend to be few such registers because of their relatively high cost in silicon area. As a result, operands must often be fetched from memory before the actual computation specified by an instruction can be completed. For most computer systems today, the amount of time it takes to access data from memory is more than 100 times the single cycle time of the CPU core. And, worse yet, the gap between typical CPU cycle times and memory-access times continues to grow. That imbalance would lead to a devastating loss in performance of most programs if there were not hardware caches in these systems. Caches hold the most frequently accessed parts of main memory in special hardware structures that have much smaller latencies than the main memory system; for example, a typical level-1 cache has an access time that is only 2-3 times slower than the single cycle time of the CPU core. They leverage a principle called *locality of reference* that characterizes common data-access patterns exhibited by most computer programs. To accommodate large working sets that do not fit in the first-level cache, many computer systems deploy a hierarchy of caches. The later levels of caches tend to be increasingly large (up to several megabytes), but as a result they also have longer access times and resulting latencies. The concept of locality is important for computer architecture, and Chapter 4 highlights the potential of exploiting locality in innovative ways.

Main memory in most modern computer systems is typically implemented with dynamic random-access memory (DRAM) chips, and it can be quite large (many gigabytes). However, it is nowhere near large enough to hold all the addressable memory space available to applications and the file systems used for long-term storage of data and programs. Therefore, nonvolatile magnetic-disk-based storage[6] is commonly used to hold this much larger collection of data. The access time for disk-based storage is several orders of magnitude larger than that of DRAM, which can expose very long delays between a request for data and the return of the data. As a result, in many computer systems, the operating system takes advantage of the situation by arranging a "context switch" to allow another pending program to run in the window of time provided by the long delay in many computer systems. Although context-switching by the operating system improves the multiprogram throughput of the overall computer system, it hurts the performance of any single application because of the associated overhead of the context-switch mechanics. Similarly, as any given program accesses other system resources, such as networking and other types of storage devices, the associated request-response delays detract from the program's ability to make use of the full peak-performance potential of the CPU core. Because each of those subsystems displays different performance characteristics, the establishment of an appropriate *system-level balance* among them is a fundamental challenge in modern computer-system design. As future technology advances improve the characteristics of the subsystems, new challenges and opportunities in balancing the overall system arise.

Today, an increasing number of computer systems deploy more than one CPU core, and this has the potential to improve system performance. In fact, there are several methods of taking advantage of the potential of *parallelism* offered by additional CPU cores, each with distinct advantages and associated challenges.

---

[6]Nonvolatile storage does not require power to retain its information. A compact disk (CD) is nonvolatile, for example, as is a computer hard drive, a USB flash key, or a book like this one.

- The first method takes advantage of the additional CPUs to improve the general *responsiveness* of the system. Instead of scheduling the execution of pending programs one at a time (as is done in single-processor systems), the operating system can schedule more than one program to run at the same time in different processors. This method tends to increase the use of the other subsystems (storage, networking, and so on), so it also demands a different system-level balance among the subsystems than do some of the other methods. From an end user's standpoint, this system organization tends to improve both the interactive responsiveness of the system and the turnaround time for any particular execution task.
- The second method takes advantage of the additional CPU cores to improve the *turnaround time* of a particular program more dramatically by running different parts of the program in parallel. This method requires programmers to use parallel-programming constructs; historically, this task has proved fairly difficult even for the most advanced programmers. In addition, such constructs tend to require particular attention to how the different parts of the program synchronize and coordinate their execution. This synchronization is a form of communication among the cooperating processors and represents a new type of overhead that detracts from exploiting the peak potential of each individual processor core. See Box 2.2 for a brief description of Amdahl's law.
- A third method takes advantage of the additional CPU cores to improve the *throughput* of a particular program. Instead of working to speed up the program's operation on a single piece of data (or dataset), the system works to increase the rate at which a collection of data (or datasets) can be processed by the program. In general, because there tends to be more independence among the collected data in this case, the development of these types of programs is somewhat easier than development of the parallel programs mentioned earlier. In addition, the degree of communication and synchronization required between the concurrently executing parts of the program tends to be much less than in the parallel-program case.

Another key aspect of modern computer systems is their ability to communicate, or network, with one another. Programmers can write programs that make use of multiple CPU cores within a single computer system or that make use of *multiple computer systems* to increase performance or to solve larger, harder problems. In those cases, it takes much longer

> **BOX 2.2**
> **Amdahl's Law**
>
> Amdahl's law sets the limit to which a parallel program can be sped up. Programs can be thought of as containing one or more parallel sections of code that can be sped up with suitably parallel hardware and a sequential section that cannot be sped up. Amdahl's law is
>
> $$\text{Speedup} = 1/[(1 - P) + P/N)],$$
>
> where P is the proportion of the code that runs in parallel and N is the number of processors.
>
> The way to think about Amdahl's law is that the faster the parallel section of the code run, the more the remaining sequential code looms as the performance bottleneck. In the limit, if the parallel section is responsible for 80 percent of the run time, and that section is sped up infinitely (so that it runs in zero time), the other 20 percent now constitutes the entire run time. It would therefore have been sped up by a factor of 5, but after that no amount of additional parallel hardware will make it go any faster.

to communicate, synchronize, and coordinate the progress of the overall program. The programs tend to break the problem into coarser-grain tasks to run in parallel, and they tend to use more explicit message-passing constructs. As a result, the development and optimization of such programs are quite different from those of the others mentioned above.

In addition to the methods described above, computer scientists are actively researching new ways to exploit multiple CPU cores, multiple computer systems, and parallelism for future systems. Considering the increased complexity of such systems, researchers are also concerned about easing the associated programming complexity exposed to the application programmer, inasmuch as programming effort has a first-order effect on time to solution of any given problem. The magnitude of these challenges and their effects on computer-system performance motivate much of this report.

TECHNOLOGY ADVANCES AND THE HISTORY OF COMPUTER PERFORMANCE

In many ways, the history of computer performance can be best understood by tracking the development and issues of the technology that underlies the machines. If one does that, an interesting pattern starts to emerge. As incumbent technologies are stretched to their practical limits, innovations are leveraged to overcome these limits. At the same time, they set the stage for a fresh round of incremental advances that eventually overtake any remaining advantages of the older technology. That technology-innovation cycle has been a driving force in the history of computer-system performance improvements.

A very early electronic computing system, called Colossus,[7] was created in 1943.[8] Its core was built with *vacuum tubes*, and although it had fairly limited utility, it ushered in the use of electronic vacuum tubes for a generation of computer systems that followed. As newer systems, such as the ENIAC, introduced larger-scale and more generalized computing, the collective power consumption of all the vacuum tubes eventually limited the ability to continue scaling the systems. In 1954, engineers at Bell Laboratories created a *discrete-transistor*-based computer system called the TRADIC.[9] Although it was not quite as fast as the fastest vacuum-tube-based systems of the day, it was much smaller and consumed much less power. More important, it heralded the era of transistor-based computer systems.[10] In 1958, Jack Kilby and Robert Noyce separately invented the *integrated circuit*, which for the first time allowed multiple transistors to be fabricated and connected on a single piece of silicon. That technology was quickly picked up by computer designers to design higher-performance and more power-efficient computer systems. This technology breakthrough inaugurated the modern computing era.

In 1965, Gordon Moore observed that the transistor density on integrated circuits was doubling with each new technology generation, and he projected that this would continue into the future.[11] (See Appendix C

---

[7]B. Jack Copeland, ed., 2006, Colossus: The Secrets of Bletchley Park's Codebreaking, New York, N.Y.: Oxford University Press.

[8]Although many types of mechanical and electromechanical computing systems were demonstrated before that, these devices were substantially limited in capabilities and deployments, so we will leave them out of this discussion.

[9]For a history of the TRADIC, see Louis C. Brown, 1999, Flyable TRADIC: The first airborne transistorized digital computer, IEEE Annals of the History of Computing 21(4): 55-61.

[10]It was not only vacuum tube power requirements that were limiting the computer industry back in the early 1060s. Packaging was a significant challenge, too—simply making all the connections needed to carry signals and power to all those tubes was seriously degrading reliability, because each connection had to be hand-soldered with some probability of failure greater than 0.0. All kinds of module packaging schemes were being tried, but none of them really solved this manufacturability problem. One of the transformative aspects of integrated circuit technology is that you get all the internal connections for free by a chemical photolithography process that not only makes them essentially free but also makes them several orders of magnitude more reliable. Were it not for that effect, all those transistors we have enjoyed ever since would be of very limited usefulness, too expensive, and too prone to failure.

[11] Gordon Moore, 1965, Cramming more components onto integrated circuits, Electronics 38(8), available online at http://download.intel.com/research/silicon/moorespaper.pdf.

for a reprint of his seminal paper.) That projection, now commonly called Moore's law, was remarkably accurate and still holds true. However, over the years, there have been some important shifts in how integrated circuits are used in computer systems. Early on, various segments of the electronics industry made use of different types of transistor devices. For high-end computer systems, the *bipolar junction transistor* (BJT) was the technology of choice. As more BJT devices were integrated into the systems, the power consumption of each chip also rose, and computer-system designers were forced to use exotic power delivery and cooling solutions. In the 1980s, another type of transistor, the *field-effect transistor* (FET), was increasingly used for smaller electronic devices, such as calculators and small computers meant for hobbyists. By the late 1980s, the power-consumption characteristics of the BJT-based computer systems hit a breaking point; around the same time, the early use of FET-based integrated circuits had demonstrated both power and cost advantages over the BJT-based technologies. Although the underlying transistors were not as fast, their characteristics enabled far greater integration potential and much lower power consumption. Today, at the heart of virtually all computer systems is a set of FET-based integrated-circuit chips.

It now appears that in some higher-end computer systems, the FET-based integrated circuits have hit their practical limits of power consumption. Although today's technologists understand how to continue increasing the level of integration (number of transistor devices) on future chips, they are not able to continue reducing the voltage or the power.[12] There are several potential new technology concepts in the research laboratories—such as carbon nanotubes, quantum dots, and biology-inspired devices—but none of them is mature enough for practical deployment. Although there is reasonable optimism that current research will eventually bring one or more new technology breakthroughs into mainstream deployment, it appears today that the technology-innovation cycle has a substantial gap that must be overcome in some other way. The industry is therefore shifting from the long-standing heritage of constantly improving the performance characteristics of single-processor-based systems (sometimes referred to as single-thread performance) to increasing the number of processors in each system. As described in the following sections, that

---

[12]The committee's emphasis on transistor performance is not intended to convey the impression that transistors are the sole determinant of computer system performance. The interconnect wiring between transistors on a chip is a first-order limiter of system clock rate and also contributes greatly to overall power dissipation. Memory and I/O systems must also scale up to avoid becoming bottlenecks to faster computer systems. The focus is on transistors here because it is possible to work around interconnect limitations (this has already been done for at least 15 years), and so far, memory and I/O have been scaling up enough to avoid being showstoppers.

puts substantial new demands and new pressures on the software side of multiprocessor-based systems.

Appendix A provides additional data on historical computer-performance trends. It illustrates that from 1985 to 2004 computer performance improved at a compound annual growth rate exceeding 50 percent, measured with the SPECint2000 and SPECfp2000 benchmarks, but after 2004 grew much more slowly.[13] Moreover, it shows that the recent slow growth is due in large part to a flattening of clock-frequency improvements needed to flatten the untenable growth in chip power requirements. The appendix closes with Kurzweil's observations on the 20th century that encourage us to seek new computer technologies.

ASSESSING PERFORMANCE WITH BENCHMARKS

As discussed earlier in this chapter, another big challenge in understanding computer-system performance is choosing the right hardware and software metrics and measurements. As this committee has already discussed, the peak-performance potential of a machine is not a particularly good metric in that the inevitable overheads associated with the use of other system-level resources and communication can diminish delivered performance substantially.

There have been innumerable efforts over the years to create benchmark suites to define a set of workloads over which to measure metrics, many of them quite successful within limited application domains. However, designing general benchmarks is difficult. Even considering hardware performance alone can be challenging because computer hardware consists of several different components (see Box 2.3). Computer systems are deployed and used in a broad variety of ways. As one might expect, different market segments have different use scenarios, and they stress the system in different ways. As a result, the appropriate benchmark to consider can vary considerably between market segments. For example,

- For casual home users, responsiveness of the GUI has high priority. The performance of the system when operating on various types of entertainment media—such as audio, video, or pictures files—is more important than it is in many other markets.
- In research settings, the computer system is an important tool for exploring and modeling ideas. As a result, the turnaround time

---

[13]SPEC benchmarks are a set of artificial workloads intended to measure a computer system's speed. A machine that achieves a SPEC benchmark score that is, say, 30 percent faster than that of another machine should feel about 30 percent faster than the other machine on real workloads.

for a given program is important because it provides the results that are an integral part of the iterative research loop directed by the researcher. That is an example of performance as time to solution. (See Box 2.4 for more on time to solution.)
- In small-business settings, the computer system tends to be used for a very wide array of applications, so high general-purpose performance is valued.
- For computer systems used in banking and other financial markets, the reliability and accuracy of the computational results, even in the face of defects or harsh external environmental conditions, are paramount. Many deployments value gross transactional throughput more than the turnaround time of any given program, except for financial-transaction turnaround time.
- In some businesses, computer systems are deployed into mission-critical roles in the overall operation of the business, for example, e-commerce-based businesses, process automation, health care, and human safety systems. In those situations, the gross reliability and "up time" characteristics of the system can be far more important than the instantaneous performance of the system at any given time.
- At the very high end, supercomputer systems tend to work on large problems with very large amounts of data. The underlying performance of the memory system can be even more important than the raw computational capability of the CPU cores involved. That can be seen as an example of throughput as performance (see Box 2.5).

Complicating matters a bit more, most computer-system deployments define some set of important physical constraints on the system. For example, in the case of a notebook-computer system, important energy-consumption and physical-size constraints must be met. Similarly, even in the largest supercomputer deployments, there are constraints on physical size, weight, power, heat, and cost. Those constraints are several orders of magnitude larger than in the notebook example, but they still are fundamental in defining the resulting performance and utility of the system. As a result, for a given market opportunity, it often makes sense to gauge the value of a computer system according to a ratio of performance to constraints. Indeed, some of the metrics most frequently used today are such ratios as performance per watt, performance per dollar, and performance per area. More generally, most computer-system customers are placing increasing emphasis on efficiency of computation rather than on gross performance metrics.

> **BOX 2.3**
> **Hardware Components**
>
> A car is not just an engine. It has a cooling system to keep the engine running efficiently and safely, an environmental system to do the same for the drivers and passengers, a suspension system to improve the ride, a transmission so that the engine's torque can be applied to the drive wheels, a radio so that the driver can listen to classic-rock stations, and cupholders and other convenience features. One might still have a useful vehicle if the radio and cupholders were missing, but the other features must be present because they all work in harmony to achieve the function of propelling the vehicle controllably and safely.
>
> Computer systems are similar. The CPU tends to get much more than its proper share of attention, but it would be useless without memory and I/O subsystems. CPUs function by fetching their instructions from memory. How did the instructions get into memory, and where did they come from? The instructions came from a file on a hard disk and traversed several buses (communication pathways) to get to memory. Many of the instructions, when executed by the CPU, cause additional memory traffic and I/O traffic. When we speak of the overall performance of a computer system, we are implicitly referring to the overall performance of all those systems operating together. For any given workload, it is common to find that one of the "links in the chain" is, in fact, the weakest link. For instance, one can write a program that only executes CPU operations on data that reside in the CPU's own register file or its internal data cache. We would refer to such a program as "CPU-bound," and it would run as fast as the CPU alone could perform it. Speeding up the memory or the I/O system would have no discernible effect on measured performance for that benchmark. Another benchmark could be written, however, that does little else but perform memory load and store operations in such a way that the CPU's internal cache is ineffective. Such a benchmark would be bound by the speed

THE INTERPLAY OF SOFTWARE AND PERFORMANCE

Although the amazing raw performance gains of the microprocessor over the last 20 years has garnered most of the attention, the overall performance and utility of computer systems are strong functions of both hardware and software. In fact, as computer systems have deployed more hardware, they have depended more and more on software technologies to harness their computational capability. Software has exploited that capability directly and indirectly. Software has directly exploited increases in computing capability by adding new features to existing software, by solving larger problems more accurately, and by solving previously unsolvable problems. It has indirectly exploited the capability through the use of abstractions in high-level programming languages, libraries, and virtual-machine execution environments. By using high-level programming languages and exploiting layers of abstraction, programmers can

of memory (and possibly by the bus that carries the traffic between the CPU and memory.) A third benchmark could be constructed that hammers on the I/O subsystem with little dependence on the speed of either the CPU or the memory.

Handling most real workloads relies on all three computer subsystems, and their performance metrics therefore reflect the combined speed of all three. Speed up only the CPU by 10 percent, and the workload is liable to speed up, but not by 10 percent—it will probably speed up in a prorated way because only the sections of the code that are CPU-bound will speed up. Likewise, speed up the memory alone, and the workload performance improves, but typically much less than the memory speedup in isolation. Numerous other pieces of computer systems make up the hardware. The CPU architectures and microarchitectures encompass instruction sets, branch-prediction algorithms, and other techniques for higher performance. Storage (disks and memory) is a central component. Memory, flash drives, traditional hard drives, and all the technical details associated with their performance (such as bandwidth, latency, caches, volatility, and bus overhead) are critical for a system's overall performance. In fact, information storage (hard-drive capacity) is understood to be increasing even faster than transistor counts on the traditional Moore's law curve,[1] but it is unknown how long this will continue. Switching and interconnect components, from switches to routers to T1 lines, are part of every level of a computer system. There are also hardware interface devices (keyboards, displays, and mice). All those pieces can contribute to what users perceive of as the "performance" of the system with which they are interacting.

---

[1] This phenomenon has been dubbed Kryder's law after Seagate executive Mark Kryder (Chip Walter, 2005, Kryder's law, Scientific American 293: 32-33, available online at http://www.scientificamerican.com/article.cfm?id=kryders-law).

---

express their algorithms more succinctly and modularly and can compose and reuse software written by others. Those high-level programming constructs make it easier for programmers to develop correct complex programs faster. Abstraction tends to trade increased human programmer productivity for reduced software performance, but the past increases in single-processor performance essentially hid much of the performance cost. Thus, modern software systems now have and rely on multiple layers of system software to execute programs. The layers can include operating systems, runtime systems, virtual machines, and compilers. They offer both an opportunity for introducing and managing parallelism and a challenge in that each layer must now also understand and exploit parallelism. The committee discusses those issues in more detail in Chapter 4 and summarizes the performance implications below.

The key performance driver to date has been *software portability*. Once

> **BOX 2.4**
> **Time To Solution**
>
> Consider a jackhammer on a city street. Assume that using a jackhammer is not a pastime enjoyable in its own right—the goal is to get a job done as soon as possible. There are a few possible avenues for improvement: try to make the jackhammer's chisel strike the pavement more times per second; make each stroke of the jackhammer more effective, perhaps by putting more power behind each stroke; or think of ways to have the jackhammer drive multiple chisels per stroke. All three possibilities have analogues in computer design, and all three have been and continue to be used. The notion of "getting the job done as soon as possible" is known in the computer industry as time to solution and has been the traditional metric of choice for system performance since computers were invented.
>
> Modern computer systems are designed according to a synchronous, pipelined schema. *Synchronous* means occurring at the same time. Synchronous digital systems are based on a system clock, a specialized timer signal that coordinates all activities in the system. Early computers had clock frequencies in the tens of kilohertz. Contemporary microprocessor designs routinely sport clocks with frequencies of over about 3-GHz range. To a first approximation, the higher the clock rate, the higher the system performance. System designers cannot pick arbitrarily high clock frequencies, however—there are limits to the speed at which the transistors and logic gates can reliably switch, limits to how quickly a signal can traverse a wire, and serious thermal power constraints that worsen in direct proportion to the clock frequency. Just as there are physical limits on how fast a jackhammer's chisel can be driven downward and then retracted for the next blow, higher computer clock rates generally yield faster time-to-solution results, but there are several immutable physical constraints on the upper limit of those clocks, and the attainable performance speedups are not always proportional to the clock-rate improvement.
>
> How much a computer system can accomplish per clock cycle varies widely from system to system and even from workload to workload in a given system. More complex computer-instruction sets, such as Intel's x86, contain instructions that intrinsically accomplish more than a simpler instruction set, such as that embodied in the ARM processor in a cell phone; but how effective the complex instructions are is a function of how well a compiler can use them. Recent

a program has been created, debugged, and put into practical use, end users' expectation is that the program not only will continue to operate correctly when they buy a new computer system but also will run faster on a new system that has been advertised as offering increased performance. More generally, once a large collection of programs have become available for a particular computing platform, the broader expectation is that they will all continue to work and speed up in later machine genera-

additions to historical instruction sets—such as Intel's SSE 1, 2, 3, and 4—attempt to accomplish more work per clock cycle by operating on grouped data that are in a compressed format (the equivalent of a jackhammer that drives multiple chisels per stroke). Substantial system-performance improvements, such as factors of 2-4, are available to workloads that happen to fit the constraints of the instruction-set extensions.

There is a special case of time-to-solution workloads: those which can be successfully sped up with dedicated hardware accelerators. Graphics processing units (GPUs)—such as those from NVIDIA, from ATI, and embedded in some Intel chipsets—are examples. These processors were designed originally to handle the demanding computational and memory bandwidth requirements of 3D graphics but more recently have evolved to include more general programmability features. With their intrinsically massive floating-point horsepower, 10 or more times higher than is available in the general-purpose (GP) microprocessor, these chips have become the execution engine of choice for some important workloads. Although GPUs are just as constrained by the exponentially rising power dissipation of modern silicon as are the GPs, GPUs are 1-2 orders of magnitude more energy-efficient for suitable workloads and can therefore accomplish much more processing within a similar power budget.

Applying multiple jackhammers to the pavement has a direct analogue in the computer industry that has recently become the primary development avenue for the hardware vendors: "multicore." The computer industry's pattern has been for the hardware makers to leverage a new silicon process technology to make a software-compatible chip that is substantially faster than any previous chips. The new, higher-performing systems are then capable of executing software workloads that would previously have been infeasible; the attractiveness of the new software drives demand for the faster hardware, and the virtuous cycle continues. A few years ago, however, thermal-power dissipation grew to the limits of what air cooling can accomplish and began to constrain the attainable system performance directly. When the power constraints threatened to diminish the generation-to-generation performance enhancements, chipmakers Intel and AMD turned away from making ever more complex microarchitectures on a single chip and began placing multiple processors on a chip instead. The new chips are called multicore chips. Current chips have several processors on a single die, and future generations will have even more.

tions. Indeed, not only has the remarkable speedup offered by industry standard (×86-compatible) microprocessors over the last 20 years forged compatibility expectation in the industry, but its success has hindered the development of alternative, noncompatible computer systems that might otherwise have kindled new and more scalable programming paradigms. As the microprocessor industry shifts to multicore processors, the rate of improvement of each individual processor is substantially diminished.

> **BOX 2.5**
> **Throughput**
>
> There is another useful performance metric besides time to solution, and the Internet has pushed it to center stage: system throughput. Consider a Web server, such as one of the machines at search giant Google. Those machines run continuously, and their work is never finished, in that new requests for service continue to arrive. For any given request for service, the user who made the request may care about time to solution, but the overall performance metric for the server is its throughput, which can be thought of informally as the number of jobs that the server can satisfy simultaneously. Throughput will determine the number and configuration of servers and hence the overall installation cost of the server "farm."
>
> Before multicore chips, the computer industry's efforts were aimed primarily at decreasing the time to solution of a system. When a given workload required the sequential execution of several million operations, a faster clock or a more capable microarchitecture would satisfy the requirement. But compilers are not generally capable of targeting multiple processors in pursuit of a single time-to-solution target; they know how to target *one* processor. Multicore chips therefore tend to be used as throughput enhancers. Each available CPU core can pop the next runnable process off the ready list, thus increasing the throughput of the system by running multiple processes concurrently. But that type of concurrency does not automatically improve the time to solution of any given process.
>
> Modern multithreading programming environments and their routine successful use in server applications hold out the promise that applying multiple threads to a single application may yet improve time to solution for multicore platforms. We do not yet know to what extent the industry's server multithreading successes will translate to other market segments, such as mobile or desktop computers. It is reasonably clear that although time-to-solution performance is topping out, throughput can be increased indefinitely. The as yet unanswered question is whether the buying public will find throughput enhancements as irresistible as they have historically found time-to-solution improvements.

The net result is that the industry is ill prepared for the rather sudden shift from ever-increasing single-processor performance to the presence of increasing numbers of processors in computer systems. (See Box 2.6 for more on instruction-set architecture compatibility and possible future outcomes.)

The reason that industry is ill prepared is that an enormous amount of existing software does not use thread-level or data-level parallelism—software did not need it to obtain performance improvements, because users simply needed to buy new hardware to get performance improvements. However, only programs that have these types of parallelism will experience improved performance in the chip multiprocessor era. Furthermore, even for applications with thread-level and data-level parallelism, it is hard to obtain improved performance with chip multiprocessor hardware because of communication costs and competition for shared resources, such as cache memory. Although expert programmers in such application domains as graphics, information retrieval, and databases have successfully exploited those types of parallelism and attained performance improvements with increasing numbers of processors, these applications are the exception rather than the rule.

Writing software that expresses the type of parallelism that hardware based on chip multiprocessors will be able to improve is the main obstacle because it requires new *software-engineering processes and tools*. The processes and tools include training programmers to solve their problems with "parallel computational thinking," new programming languages that ease the expression of parallelism, and a new software stack that can exploit and map the parallelism to hardware that is evolving. Indeed, the outlook for overcoming this obstacle and the ability of academics and industry to do it are primary subjects of this report.

THE ECONOMICS OF COMPUTER PERFORMANCE

There should be little doubt that computers have become an indispensable tool in a broad array of businesses, industries, research endeavors, and educational institutions. They have enabled profound improvement in automation, data analysis, communication, entertainment, and personal productivity. In return, those advances have created a *virtuous economic cycle* in the development of new technologies and more advanced computing systems. To understand the sustainability of continuing improvements in computer performance, it is important first to understand the health of this cycle, which is a critical economic underpinning of the computer industry.

From a purely technological standpoint, the engineering community has proved to be remarkably innovative in finding ways to continue to reduce microelectronic feature sizes. First, of course, industry has integrated more and more transistors into the chips that make up the computer systems. Fortunate side effects are improvements in speed and power efficiency of the individual transistors. Computer architects have learned to make use of the increasing numbers and improved characteristics of the transistors to design continually higher-performance computer systems. The demand for the increasingly powerful computer systems has generated sufficient revenue to fuel the development of the next round of technology while providing profits for the companies leading the charge. Those relationships form the basis of the virtuous economic

> **BOX 2.6**
> **Instruction-Set Architecture: Compatibility**
>
> This history of computing hardware has been dominated by a few franchises. IBM first noticed the trend of increasing performance in the 1960s and took advantage of it with the System/360 architecture. That instruction-set architecture became so successful that it motivated many other companies to make computer systems that would run the same software codes as the IBM System/360 machines; that is, they were building instruction-set-*compatible* computers. The value of that approach is clearest from the end user's perspective—compatible systems worked as expected "right out of the box," with no recompilation, no alterations to source code, and no tracking down of software bugs that may have been exposed by the process of migrating the code to a new architecture and toolset.
>
> With the rise of personal computing in the 1980s, compatibility has come to mean the degree of compliance with the Intel architecture (also known as IA-32 or x86). Intel and other semiconductor companies, such as AMD, have managed to find ways to remain compatible with code for earlier generations of x86 processors. That compatibility comes at a price. For example, the floating-point registers in the x86 architecture are organized as a stack, not as a randomly accessible register set, as all integer registers are. In the 1980s, stacking the floating-point registers may have seemed like a good idea that would benefit compiler writers; but in 2008, that stack is a hindrance to performance, and x86-compatible chips therefore expend many transistors to give the architecturally required appearance of a stacked floating-point register set—only to spend more transistors "under the hood" to undo the stack so that modern performance techniques can be applied. IA-32's instruction-set encoding and its segmented addressing scheme are other examples of old baggage that constitute a tax on every new x86 chip.
>
> There was a time in the industry when much architecture research was expended on the notion that because every new compatible generation of chips must carry the aggregated baggage of its past and add ideas to the architecture to keep it current, surely the architecture would eventually fail of its own accord, a victim of its own success. But that has not happened. The baggage is there, but the magic of Moore's law is that so many additional transistors are made available in each new generation, that there have always been enough to reimplement the baggage and to incorporate enough innovation to stay competitive. Over time, such non-x86-compatible but worthy competitors as DEC's Alpha, SGI's MIPS, Sun's SPARC, and the Motorola/IBM PowerPC architectures either have found a niche in market segments, such as cell phones or other embedded products, or have disappeared.

and technology-advancement cycles that have been key underlying drivers in the computer-systems industry over the last 30 years.

There are many important applications of semiconductor technology beyond the desire to build faster and faster high-end computer systems. In particular, the electronics industry has leveraged the advances

The strength of the x86 architecture was most dramatically demonstrated when Intel, the original and major supplier of x86 processors, decided to introduce a new, non-x86 architecture during the transition from 32-bit to 64-bit addressing in the 1990s. Around the same time, however, AMD created a processor with 64-bit addressing compatible with the x86 architecture, and customers, again driven by the existence of the large software base, preferred the 64-bit x86 processor from AMD over the new IA-64 processor from Intel. In the end, Intel also developed a 64-bit x86-compatible processor that is now far outselling its IA-64 (Itanium) processor.

With the rise of the cell phone and other portable media and computing appliances, yet another dominant architectural approach has emerged: the ARM architecture. The rapidly growing software base for portable applications running on ARM processors has made the compatible series of processors licensed by ARM the dominant processors for embedded and portable applications. As seen in the dominance of the System/360 architecture for mainframe computers, x86 for personal computers and networked servers, and the ARM architecture for portable appliances, there will be an opportunity for a new architecture or architectures as the industry moves to multicore, parallel computing systems. Initial moves to chip-multiprocessor systems are being made with existing architectures based primarily on the x86. The computing industry has accumulated a lot of history on this subject, and it appears safe to say that the era in which compatibility is an absolute requirement will probably end not because an incompatible but compellingly faster competitor has appeared but only when one of the following conditions takes hold:

- Software translation (automatic conversion from code compiled for one architecture to be suitable for running on another) becomes ubiquitous and so successful that strict hardware compatibility is no longer necessary for the user to reap the historical benefits.
- Multicore performance has "topped out" to the point where most buyers no longer perceive enough benefit to justify buying a new machine to replace an existing, still-working one.
- The fundamental hardware-software business changes so substantially that the whole idea of compatibility is no longer relevant. Interpreted and dynamically compiled languages—such as Java, PHP, and JavaScript ("write once run anywhere")—are harbingers of this new era. Although their performance overhead is sometimes enough for the performance advantages of compiled code to outweigh programmer productivity, JavaScript and PHP are fast becoming the languages of choice on the client side and server side, respectively, for Web applications.

to create a wide variety of new form-factor devices (notebook computers, smart phones, and GPS receivers, to name just a few). Although each of those devices tends to have more substantial constraints (size, power, and cost) than traditional computer systems, they often embody the computational and networking capabilities of previous generations of higher-end computer systems. In light of the capabilities of the smaller form-factor devices, they will probably play an important role in unleashing the aggregate performance potential of larger-scale networked systems in the future. Those additional market opportunities have strong economic underpinnings of their own, and they have clearly reaped benefits from deploying technology advances driven into place by the computer-systems industry. In many ways, the incredible utility of computing not only has provided direct improvement in productivity in many industries but also has set the stage for amazing growth in a wide array of codependent products and industries.

In recent years, however, we have seen some potentially troublesome changes in the traditional return on investment embedded in this virtuous cycle. As we approach more of the fundamental physical limits of technology, we continue to see dramatic increases in the costs associated with technology development and in the capital required to build fabrication facilities to the point where only a few companies have the wherewithal even to consider building these facilities. At the same time, although we can pack more and more transistors into a given area of silicon, we are seeing diminishing improvements in transistor performance and power efficiency. As a result, computer architects can no longer rely on those sorts of improvements as means of building better computer systems and now must rely much more exclusively on making use of the increased transistor-integration capabilities.

Our progress in identifying and meeting the broader value propositions has been somewhat mixed. On the one hand, multiple processor cores and other system-level features are being integrated into monolithic pieces of silicon. On the other hand, to realize the benefits of the multiprocessor machines, the software that runs on them must be conceived and written in a different way from what most programmers are accustomed to. From an end-user perspective, the hardware and the software must combine seamlessly to offer increased value. It is increasingly clear that the computer-systems industry needs to address those software and programmability concerns or risk the ability to offer the next round of compelling customer value. Without performance incentives to buy the next generation of hardware, the economic virtuous cycle is likely to break down, and this would have widespread negative consequences for many industries.

In summary, the sustained viability of the computer-systems industry is heavily influenced by an underlying virtuous cycle that connects continuing customer perception of value, financial investments, and new products getting to market quickly. Although one of the primary indicators of value has traditionally been the ever-increasing performance of each individual compute node, the next round of technology improvements on the horizon will not automatically enhance that value. As a result, many computer systems under development are betting on the ability to exploit multiple processors and alternative forms of parallelism in place of the traditional increases in the performance of individual computing nodes. To make good on that bet, there need to be substantial breakthroughs in the software-engineering processes that enable the new types of computer systems. Moreover, attention will probably be focused on high-level performance issues in large systems at the expense of time to market and the efficiency of the virtuous cycle.

3

Power Is Now Limiting Growth in Computing Performance

The previous chapters define computer performance and discuss why its continued growth is critical for the U.S. economy. This chapter explores the relationship between computer performance and power consumption. The limitations imposed by power consumption are responsible for the present movement toward parallel computation. This chapter argues that such limitations will constrain the computation performance of even parallel computing systems unless computer designers take fundamentally different approaches.

The laws of thermodynamics are evident to anyone who has ever used a laptop computer on his or her lap: the computer becomes hot. When you run a marathon, your body is converting an internal supply of energy formed from the food you eat into two outputs: mechanical forces produced by your body's muscles and heat. When you drive your car, the engine converts the energy stored in the gasoline into the kinetic energy of the wheels and vehicle motion and heat. If you put your hand on an illuminated light bulb, you discover that the bulb is not only radiating the desired light but also generating heat. Heat is the unwanted, but inevitable, side effect of using energy to accomplish any physical task—it is not possible to convert all the input energy perfectly into the desired results without wasting some energy as heat.

A vendor who describes a "powerful computer" is trying to characterize the machine as fast, not thermally hot. In this report, the committee uses *power* to refer to the use of electric energy and *performance* to mean computational capability. When we refer to power in the context of a computer system, we are talking about energy flows: the rate at which energy must be provided to the computer system from a battery or wall outlet, which is the same as the rate at which that energy, now converted to heat, must be extracted from the system. The temperature of the chip or system rises above the ambient temperature and causes heat energy to flow into the environment. To limit the system's temperature rise, heat must be extracted efficiently. (Transferring heat from the computer system to the environment is the task of the cooling subsystem in a computer.) Thus, referring to a chip's power requirements is equivalent to talking about power consumed and dissipated.

When we talk about scaling computing performance, we implicitly mean to increase the computing performance that we can buy for each dollar we spend. If we cannot scale down the energy per function as fast as we scale up the performance (functions per second), the power (energy per second) consumed by the system will rise, and the increase in power consumption will increase the cost of the system. More expensive hardware will be needed to supply the extra power and then remove the heat that it generates. The cost of managing the power in and out of the system will rise to dominate the cost of the hardware.

Historically, technology scaling has done a good job of scaling down energy cost per function as the total cost per function dropped, and so the overall power needs of systems were relatively constant as performance (functions per second) dramatically increased. Recently, the cost per function has been dropping faster than the power per function, which means that the overall power of constant-cost chips has been growing. The power problem is getting worse because of the recent difficulty in continuing to scale down power-supply voltages, as is described later in this chapter.

Our ability to supply power and cool chips is not improving rapidly, so for many computers the performance per dollar is now limited by power issues. In addition, computers are increasingly available in a variety of form-factors and many, such as cell phones, have strict power limits because of user constraints. People do not want to hold hot cell phones, and so the total power budget needs to be under a few watts when the phone is active. Designers today must therefore find the best performance that can be achieved within a specified power envelope.

To assist in understanding these issues, this chapter reviews integrated circuit (IC) technology scaling. It assumes general knowledge of electric circuits, and some readers may choose to review the findings listed here and then move directly to Chapter 4. The basic conclusions of this chapter are as follows:

- Power consumption has become the limiting constraint on future growth in single-processor performance.
- Power limitations on individual processors have resulted in chips that have multiple lower-power processors that, in some applications, yield higher aggregate performance than chips with single power-limited processors.
- Even as the computing industry successfully shifts to multiple, simpler, and lower-power processor cores per chip, it will again face performance limits within a few years because of aggregate, full-chip power limitations.
- Like complementary metal oxide semiconductor (CMOS) transistors, many of the electronic devices being developed today as potential replacements for CMOS transistors are based on regulating the flow of electrons over a thermodynamic barrier and will face their own power limits.
- A promising approach to enabling more power-efficient computation is to design application-specific or algorithm-specific computational units. For that approach to succeed, new chip design and verification methods (such as advanced electronic-design automation tools) will need to be developed to reduce the time and cost of IC design, and new IC fabrication methods will be needed to reduced the one-time mask costs.
- The present move to chip multiprocessors is a step in the direction of using parallelism to sustain growth in computing performance. However, this or any other hardware architecture can succeed only if appropriate software can be developed to exploit the parallelism in the hardware effectively. That software challenge is the subject of Chapter 4.

For intrepid readers prepared to continue with this chapter, the committee starts by explaining how classic scaling enabled the creation of cheaper, faster, and lower-power circuits; in essence, scaling is responsible for Moore's law. The chapter also discusses why modern scaling produces smaller power gains than before. With that background in technology scaling, the chapter then explains how computer designers have used improving technology to create faster computers. The discussion highlights why processor performance grew faster than power efficiency and why the problem is more critical today. The next sections explain the move to chips that have multiple processors and clarify both the power-efficiency advantage of parallel computing and the limitations of this approach. The chapter concludes by mentioning some alternative technologies to assess the potential advantages and the practicality of these approaches. Alternatives to general-purpose processors are examined as a way to address power limitations. In short, although incremental advances in computing performance will continue, overcoming the power constraint is difficult or potentially impossible and will require radical rethinking of computation and of the logic gates used to build computing systems.

BASIC TECHNOLOGY SCALING

Although this report focuses on computers based on IC processors, it is useful to remember that computers have historically used various technologies. The earliest electric computers were built in the 1940s and used mechanical relays.[1,2] Vacuum tubes enabled faster electronic computers. By the 1960s, the technology for building computers changed again, to transistors that were smaller and had lower cost, lower power, and greater reliability. Within a decade, computers migrated to ICs instead of discrete transistors and were able to scale up performance as the technology scaled down the size of transistors in the ICs. Each technology change, from relays to vacuum tubes to transistors to ICs, decreased the cost, increased the performance of each function, and decreased the energy per function. Those three factors enabled designers to continue to build more capable computers for the same cost.

Early IC computers were built with bipolar transistors[3] in their ICs, which offered high performance but used relatively high power (compared with other IC options). By the late 1970s, low-end computers used NMOS[4] technology, which offered greater density and thus lower cost per function but also lower-speed gates than the bipolar alternative. As scaling continued, the cost per function was dropping rapidly, but the energy needs of each gate were not dropping as rapidly, so the power-dissipation

---

[1] Raúl Rojas, 1997, Konrad Zuse's legacy: The architecture of the Z1 and Z3, IEEE Annals of the History of Computing 19(2): 5-19.

[2] IBM, 2010, Feeds, speeds and specifications, IBM Archives, website, available online at http://www-03.ibm.com/ibm/history/exhibits/markI/markI_feeds.html.

[3] Silicon ICs use one of two basic structures for building switches and amplifiers. Both transistor structures modify the silicon by adding impurities to it that increase the concentration of electric carriers—electrons for N regions and holes for P regions—and create three regions: two Ns separated by a P or two Ps separated by an N. That the electrons are blocked by holes (or vice versa) means that there is little current flow in all these structures. The first ICs used NPN bipolar transistors, in which the layers are formed vertically in the material and the current flow is a bulk property that requires electrons to flow into the P region (the base) and holes to flow into the top N region (the emitter).

[4] NMOS transistors are lateral devices that work by having a "gate" terminal that controls the surface current flow between the "source" and "drain" contacts. The source-drain terminals are doped N and supply the electrons that flow through a channel; hence the name NMOS. Doping refers to introducing impurities to affect the electrical properties of the semiconductor. PMOS transistors make the source and drain material P, so holes (electron deficiencies) flow across the channels.

requirements of the chips were growing. By the middle 1980s, most processor designers moved from bipolar and NMOS to CMOS[5] technology. CMOS gates were slower than those of NMOS or bipolar circuits but dissipated much less energy, as described in the section below. Using CMOS technology reduced the energy per function by over an order of magnitude and scaled well. The remainder of this chapter describes CMOS technology, its properties, its limitations, and how it affects the potential for growth in computing performance.

CLASSIC CMOS SCALING

Computer-chip designers have used the scaling of feature sizes (that is, the phenomenon wherein the same functionality requires less space on a new chip) to build more capable, more complex devices, but the resulting chips must still operate within the power constraints of the system. Early chips used circuit forms (bipolar or NMOS circuits) that dissipated power all the time, whether the gate[6] was computing a new value or just holding the last value. Even though scaling allowed a decrease in the power needed per gate, the number of gates on a chip was increasing faster than the power requirements were falling; by the early to middle 1980s, chip power was becoming a design challenge. Advanced chips were dissipating many watts;[7] one chip, the HP Focus processor, for example, was dissipating over 7 W, which at the time was a very large number.[8]

Fortunately, there was a circuit solution to the problem. It became possible to build a type of gate that dissipated power only when the output value changed. If the inputs were stable, the circuit would dissipate practically no power. Furthermore, the gate dissipated power only as long as it took to get the output to transition to its new value and then returned to a zero-power state. During the transition, the gate's power requirement was comparable with those of the previous types of gates, but because the transition lasts only a short time, even in a very active machine a gate

---

[5]The C in CMOS stands for complementary. CMOS uses both NMOS and PMOS transistors.

[6]A logic gate is a fundamental building block of a system. Gates typically have two to four inputs and produce one input. These circuits are called logic gates because they compute simple functions used in logic. For example, an AND gate takes two inputs (either 1s or 0s) and returns 1 if both are 1s and 0 if either is 0. A NOT gate has only one input and returns 1 if the input is 0 and 0 if the input is 1.

[7]Robert M. Supnick, 1984, MicroVAX 32, a 32 bit microprocessor, IEEE Journal of Solid State Circuits 19(5): 675-681, available online at http://ieeexplore.ieee.org/stamp/stamp.jsp?arnumber=1052207&isnumber=22598.

[8]Joseph W. Beyers, Louis J. Dohse, Jospeh P. Fucetola, Richard L. Kochis, Cliffird G. Lob, Gary L. Taylor, and E.R. Zeller, 1981, A 32-bit VLSI CPU chip, IEEE Journal of Solid-State Circuits 16(5): 537-542, available online at http://ieeexplore.ieee.org/stamp/stamp.jsp?arnumber=1051634&isnumber=22579.

TABLE 3.1 Scaling Results for Circuit Performance

Device or Circuit Parameter	Scaling Factor
Device dimension t_{ox}, L, W	$1/k$
Doping concentration N_a	k
Voltage V	$1/k$
Current I	$1/k$
Capacitance eA/t	$1/k$
Delay time per circuit VC/I	$1/k$
Power dissipation per circuit VI	$1/k^2$
Power density VI/A	1

SOURCE: Reprinted from Robert H. Dennard, Fritz H. Gaensslen, Hwa-Nien. Yu, V. Leo Rideout, Ernest Bassous, and Andre R. LeBlanc, 1974, Design of ion-implanted MOSFETS with very small physical dimensions, IEEE Journal of Solid State Circuits 9(5): 256-268.

would be in transition around 1 percent of the time. Thus, moving to the new circuit style decreased the power consumed by a computation by a factor of over 30.[9] The new circuit style was called complementary MOS, or CMOS.

A further advantage of CMOS gates was that their performance and power were completely determined by the MOS transistor properties. In a classic 1974 paper, reprinted in Appendix D, Robert Dennard et al. showed that the MOS transistor has a set of very convenient scaling properties.[10] The scaling properties are shown in Table 3.1, taken from that paper. If all the voltages in a MOS device are scaled down with the physical dimensions, the operation of the device scales in a particularly favorable way. The gates clearly become smaller because linear dimensions are scaled. That scaling also causes gates to become faster with lower energy per transition. If all dimensions and voltages are scaled by the scaling factor κ (κ has typically been 1.4), after scaling the gates become $(1/\kappa)^2$ their previous size, and κ^2 more gates can be placed on a chip of roughly the same size and cost as before. The delay of the gate also decreases by $1/\kappa$, and, most important, the energy dissipated each time the gate switches decreases by $(1/\kappa)^3$. To understand why the energy drops so rapidly, note that the energy that the gate dissipates is proportional to the energy that is stored at the output of the gate. That energy is proportional to a quan-

---

[9] The old style dissipated power only half the time; this is why the improvement was by a factor of roughly 30.

[10] Robert H. Dennard, Fritz H. Gaensslen, Hwa-Nien. Yu, V. Leo Rideout, Ernest Bassous, and Andre R. LeBlanc, 1974, Design of ion-implanted MOSFETS with very small physical dimensions, IEEE Journal of Solid State Circuits 9(5):256–268.

tity called capacitance[11] and the square of the supply voltage. The load capacitance of the wiring decreases by $1/\kappa$ because the smaller gates make all the wires shorter and capacitance is proportional to length. Therefore, the power requirements per unit of space on the chip (mm²), or energy per second per mm², remain constant:

$$\text{Power} = (\text{number of gates})(C_{Load/gate})(\text{Clock Rate})(V_{supply}^2)$$
$$\text{Power density} = N_g C_{load} F_{clk} V_{dd}^2$$

N_g = CMOS gates per unit area
C_{load} = capacitive load per CMOS gate
F_{clk} = clock frequency
V_{dd} = supply voltage

$$\text{Power density} = (\kappa^2)(1/\kappa)(\kappa)(1/\kappa)^2 = 1$$

That the power density (power requirements per unit space on the chip, even when each unit space contains many, many more gates) can remain constant across generations of CMOS scaling has been a critical property underlying progress in microprocessors and in ICs in general. In every technology generation, ICs can double in complexity and increase in clock frequency while consuming the same power and not increasing in cost.

Given that description of classic CMOS scaling, one would expect the power of processors to have remained constant since the CMOS transition, but this has not been the case. During the late 1980s and early 1990s, supply voltages were stuck at 5 V for system reasons. So power density would have been expected to increase as technology scaled from 2 μm to 0.5 μm. However, until recently supply voltage has scaled with technology, but power densities continued to increase. The cause of the discrepancy is explained in the next section. Note that Figure 3.1 shows no microprocessors above about 130 W; this is because 130 W is the physical limit for air cooling, and even approaching 130 W requires massive heat sinks and local fans.

---

[11]Capacitance is a measure of how much electric charge is needed to increase the voltage between two points and is also the proportionality constant between energy stored on a wire and its voltage. Larger capacitors require more charge (and hence more current) to reach a voltage than a smaller capacitor. Physically larger capacitors tend to have larger capacitance. Because all wires have at least some parasitic capacitance, even just signaling across the internal wires of a chip dissipates some power. Worse, to minimize the time wasted in charging or discharging, the transistors that drive the signal must be made physically larger, and this increases their capacitance load, which the prior gate must drive, and costs power and increases the incremental die size.

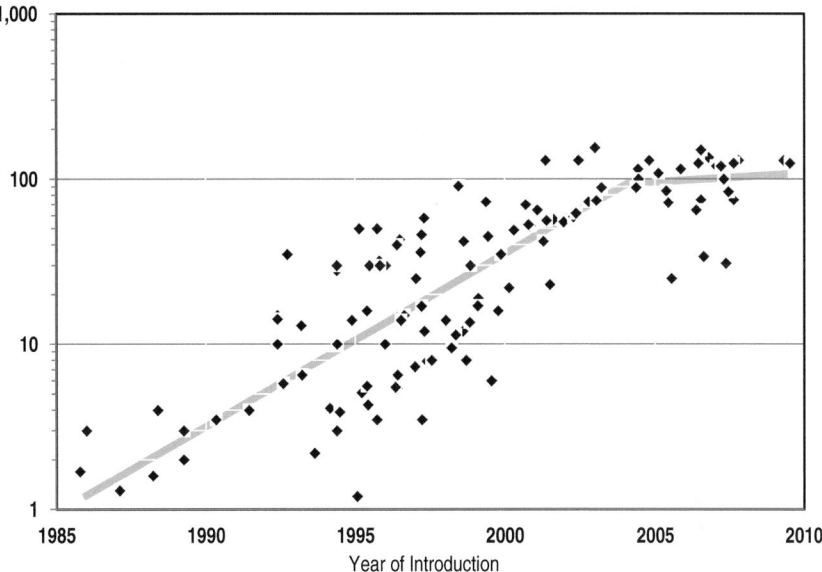

FIGURE 3.1 Microprocessor power dissipation (watts) over time (1985-2010).

HOW CMOS-PROCESSOR PERFORMANCE IMPROVED EXPONENTIALLY, AND THEN SLOWED

Microprocessor performance, as measured against the SPEC2006 benchmark[12,13,14] was growing exponentially at the rate of more than 50 percent per year (see Figure 3.2). That phenomenal single-processor performance growth continued for 16 years and then slowed substantially[15] partially because of power constraints. This section briefly describes how those performance improvements were achieved and what contributed to the slowdown in improvement early in the 2000s.

To achieve exponential performance growth, microprocessor designers scaled processor-clock frequency and exploited *instruction-level paral-*

---

[12] For older processors, SPEC2006 numbers were estimated from older versions of the SPEC benchmark by using scaling factors.

[13] John L. Henning, 2006, SPEC CPU2006 benchmark descriptions, ACM SIGARCH Computer Architecture News 34(4): 1-17.

[14] John L. Henning, 2007, SPEC CPU suite growth: An historical perspective, ACM SIGARCH Computer Architecture News 35(1): 65-68.

[15] John L. Hennessy and David A. Patterson, 2006, Computer Architecture: A Quantitative Approach, fourth edition, San Francisco, Cal.: Morgan Kauffman, pp. 2-4.

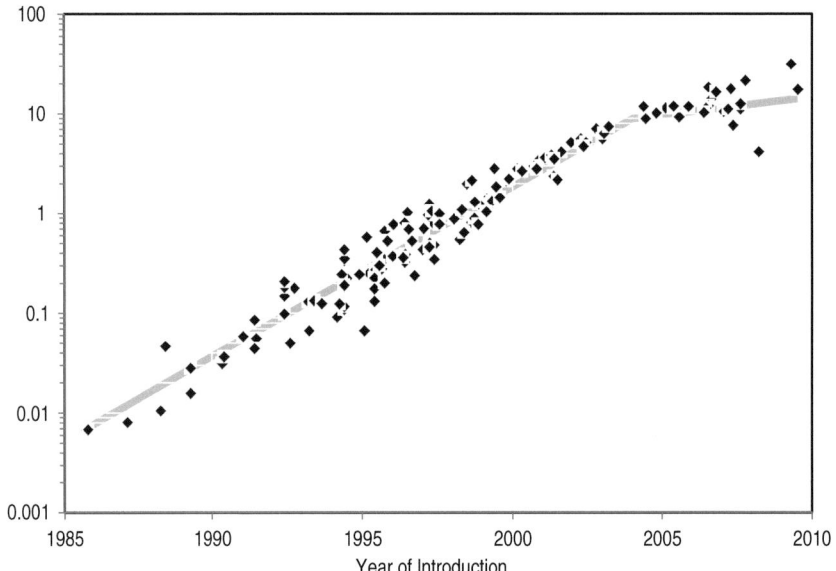

FIGURE 3.2 Integer application performance (SPECint2006) over time (1985-2010).

lelism (ILP) to increase the number of *instructions per cycle*.[16,17,18] The power problem arose primarily because clock frequencies were increasing faster than the basic assumption in Dennard scaling (described previously). The assumption there is that clock frequency will increase inversely proportionally to the basic gate speed. But the increases in clock frequency were made because of improvements in transistor speed due to CMOS-technology scaling combined with improved circuits and architecture. The designs also included deeper pipelining and required less logic (fewer operations on gates) per pipeline stage.[19]

Separating the effect of technology scaling from those of the other improvements requires examination of metrics that depend solely on the improvements in underlying CMOS technology (and not other improve-

---

[16]Ibid.

[17]Mark Horowitz and William Dally, 2004, How scaling will change processor architecture, IEEE International Solid States Circuits Conference Digest of Technical Papers, San Francisco, Cal., February 15-19, 2004, pp. 132-133.

[18]Vikas Agarwal, Stephen W. Keckler, and Doug Burger, 2000, Clock rate versus IPC: The end of the road for conventional microarchitectures, Proceedings of the 27th International Symposium Computer Architecture, Vancouver, British Columbia, Canada, June 12-14, 2000, pp. 248-259.

[19]Pipelining is a technique in which the structure of a processor is partitioned into simpler, sequential blocks. Instructions are then executed in assembly-line fashion by the processor.

ments in circuits and architecture). (See Box 3.1 for a brief discussion of this separation.) Another contribution to increasing power requirements per chip has been the nonideal scaling of interconnecting wires between CMOS devices. As the complexity of computer chips increased, it was not sufficient simply to place two copies of the previous design on the new chip. To yield the needed performance improvements, new commu-

---

BOX 3.1
Separating the Effects of CMOS Technology Scaling on Performance by Using the FO4 Metric

To separate the effect of CMOS technology scaling from other sorts of optimizations, processor clock-cycle time can be characterized by using the technology-dependent delay metric *fanout-of-four delay* (FO4), which is defined as the delay of one inverter driving four copies of an equally sized inverter.[1,2] The metric measures the clock cycle in terms of the basic gate speed and gives a number that is relatively technology-independent. In Dennard scaling, FO4/cycle would be constant. As it turns out, clock-cycle time decreased from 60-90 FO4 at the end of the 1980s to 12-25 in 2003-2004. The increase in frequency caused power to increase and, combined with growing die size, accounted for most of the power growth until the early 2000s.

That fast growth in clock rate has stopped, and in the most recent machines the number of FO4 in a clock cycle has begun to increase. Squeezing cycle time further does not result in substantial performance improvements, but it does increase power dissipation, complexity, and cost of design.[3,4] As a result, clock frequency is not increasing as fast as before (see Figure 3.3). The decrease in the rate of growth in of clock frequency is also forecast in the 2009 ITRS semiconductor roadmap,[5] which shows the clock rate for the highest-performance single processors no more than doubling each decade over the foreseeable future.

---

[1]David Harris, Ron Ho, Gu-Yeon Wei, and Mark Horowitz, The fanout-of-4 inverter delay metric, Unpublished manuscript, May 29, 2009, available online at http://www-vlsi.stanford.edu/papers/dh_vlsi_97.pdf.
[2]David Harris and Mark Horowitz, 1997, Skew-tolerant Domino circuits, IEEE Journal of Solid-State Circuits 32(11): 1702-1711.
[3]Mark Horowitz and William Dally, 2004, How scaling will change processor architecture, IEEE International Solid States Circuits Conference Digest of Technical Papers, San Francisco, Cal., February 15-19, 2004, pp. 132-133.
[4]Vikas Agarwal, Stephen W. Keckler, and Doug Burger, 2000, Clock rate versus IPC: The end of the road for conventional microarchitectures. Proceedings of the 27th International Symposium on Computer Architecture, Vancouver, British Columbia, Canada, June 12-14, 2000, pp. 248-259.
[5]See http://www.itrs.net/Links/2009ITRS/Home2009.htm.

nication paths across the entire machine were needed—interconnections that did not exist in the previous generation. To provide the increased interconnection, it was necessary to increase the number of levels of metal interconnection available on a chip, and this increased the total load capacitance faster than assumed in Dennard scaling. Another factor that has led to increases in load capacitance is the slight scaling up of wire capacitance per length. That has been due to increasing side-to-side capacitance because practical considerations limited the amount of vertical scaling possible in wires. Technologists have attacked both those issues by creating new insulating materials that had lower capacitance per length (known as low K dielectrics); this has helped to alleviate the problem, but it continues to be a factor in shrinking technologies.

One reason that increasing clock rate was pushed so hard in the 1990s, apart from competitive considerations in the chip market, was that finding parallelism in an application constructed from a sequential stream of instructions (ILP) was difficult, required large hardware structures, and was increasingly inefficient. Doubling the hardware (number of transistors available) generated only about a 50 percent increase in performance—a relationship that at Intel was referred to as Pollack's rule.[20] To continue to scale performance required dramatic increases in clock frequency, which drove processor power requirements. By the early 2000s, processors had attained power dissipation levels that were becoming difficult to handle cheaply, so processor power started to level out. Consequently, single-processor performance improvements began to slow. The upshot is a core finding and driver of the present report (see Figure 3.3), namely,

Finding: After many decades of dramatic exponential growth, single-processor performance is increasing at a much lower rate, and this situation is not expected to improve in the foreseeable future.

HOW CHIP MULTIPROCESSORS ALLOW SOME CONTINUED PERFORMANCE-SCALING

One way around the performance-scaling dilemma described in the previous section is to construct computing systems that have multiple, explicitly parallel processors. For parallel applications, that arrangement should get around Pollack's rule; doubling the area should double the

---

[20]Patrick P. Gelsinger, 2001, Microprocessors for the new millennium: Challenges, opportunities, and new frontiers, IEEE International Solid-State Circuits Conference Digest of Technical Papers, San Francisco, Cal., February 5-7, 2001, pp. 22-25. Available online at http://ieeexplore.ieee.org/stamp/stamp.jsp?arnumber=912412&isnumber=19686.

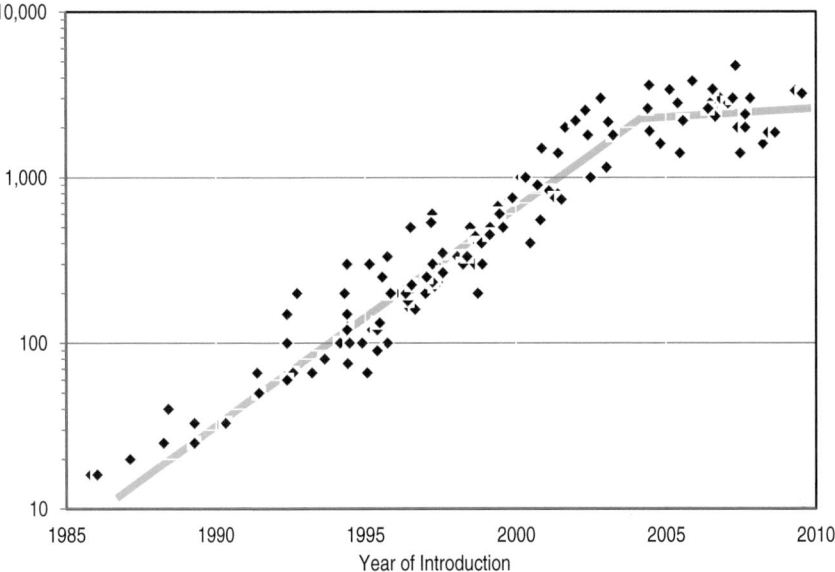

FIGURE 3.3 Microprocessor-clock frequency (MHz) over time (1985-2010).

expected performance. One might think that it should therefore be possible to continue to scale performance by doubling the number of processor cores. And, in fact, since the middle 1990s, some researchers have argued that *chip multiprocessors* (CMPs) can exploit capabilities of CMOS technology more effectively than single-processor chips.[21] However, during the 1990s, the performance of single processors continued to scale at the rate of more than 50 percent per year, and power dissipation was still not a limiting factor, so those efforts did not receive wide attention. As single-processor performance scaling slowed down and the air-cooling power-dissipation limit became a major design constraint, researchers and industry shifted toward CMPs or *multicore* microprocessors.[22]

---

[21] Kunle Olukotun, Basem A. Nayfeh, Lance Hammond, Ken Wilson, and Kunyung Chang, 1996, The case for a single-chip multiprocessor, Proceedings of 7th International Conference on Architectural Support for Programming Languages and Operating Systems, Cambridge, Mass., October 1-5, 1996, pp. 2-11.

[22] Listed here are some of the references that document, describe, and analyze this shift: Michael Bedford Taylor, Walter Lee, Jason Miller, David Wentzlaff, Ian Bratt, Ben Greenwald, Henry Hoffmann, Paul Johnson, Jason Kim, James Psota, Arvind Saraf, Nathan Shnidman, Volker Strumpen, Matt Frank, Saman Amarasinghe, and Anant Agarwal, 2004, Evaluation of the raw microprocessor: An exposed-wire-delay architecture for ILP and streams, Proceedings of the 31st Annual International Symposium on Computer Architecture, Munich,

The key observation motivating a CMP design is that to increase performance when the overall design is power-limited, each instruction needs to be executed with less energy. The power consumed is the energy per instruction times the performance (instructions per second). Examination of Intel microprocessor-design data from the i486 to the Pentium 4, for example, showed that power dissipation scales as performance raised to the 1.73 power after technology improvements are factored out. If the energy per instruction were constant, the relationship should be linear. Thus, the Intel Pentium 4 is about 6 times faster than the i486 in the same

---

Germany, June 19-23, 2004, pp. 2-13; Jung Ho Ahn, William J. Dally, Brucek Khailany, Ujval J. Kapasi, and Abhishek Das, 2004, Evaluating the imagine stream architecture, Proceedings of the 31st Annual International Symposium on Computer Architecture, Munich, Germany, June 19-23, 2004, pp. 14-25; Brucek Khailany, Ted Williams, Jim Lin, Eileen Peters Long, Mark Rygh, Deforest W. Tovey, and William Dally, 2008, A programmable 512 GOPS stream processor for signal, image, and video processing, IEEE Journal of Solid-State Circuits 43(1): 202-213; Christoforos Kozyrakis and David Patterson, 2002, Vector vs superscalar and VLIW architectures for embedded multimedia benchmarks, Proceedings of the 35th Annual ACM/IEEE International Symposium on Microarchitecture, Istanbul, Turkey, November 18-22, 2002, pp. 283-293; Luiz André Barroso, Kourosh Gharachorloo, Robert McNamara, Andreas Nowatzyk, Shaz Qadeer, Barton Sano, Scott Smith, Robert Stets, and Ben Verghese, 2000, Piranha: A scalable architecture based on single-chip multiprocessing, Proceedings of the 27th Annual International Symposium on Computer Architecture, Vancouver, British Columbia, Canada, June 10-14, 2000, pp. 282-293; Poonacha Kongetira, Kathirgamar Aingaran, and Kunle Olukotun, 2005, "Niagara: A 32-way multithreaded SPARC processor, IEEE Micro 25(2): 21-29; Dac C. Pham, Shigehiro Asano, Mark D. Bolliger, Michael N. Day, H. Peter Hofstee, Charles Johns, James A. Kahle, Atsushi Kameyama, John Keaty, Yoshio Masubuchi, Mack W. Riley, David Shippy, Daniel Stasiak, Masakazu Suzuoki, Michael F. Wang, James Warnock, Stephen Weitzel, Dieter F. Wendel, Takeshi Yamazaki, and Kazuaki Yazawa, 2005, The design and implementation of a first-generation CELL processor, IEEE International Solid-State Circuits Conference Digest of Technical Papers, San Francisco, Cal., February 10, 2005, pp. 184-185; R. Kalla, B. Sinharoy, and J.M. Tendler, 2004, IBM POWER5 chip: A dual-core multithreaded processor, IEEE Micro Magazine 24(2): 40-47; Toshinari Takayanagi, Jinuk Luke Shin, Bruce Petrick, Jeffrey Su, and Ana Sonia Leon, 2004, A dual-core 64b UltraSPARC microprocessor for dense server applications, IEEE International Solid-State Circuits Conference Digest of Technical Papers, San Francisco, Cal., February 15-19, 2004, pp. 58-59; Nabeel Sakran, Marcelo Uffe, Moty Mehelel, Jack Dowweck, Ernest Knoll, and Avi Kovacks, 2007, The implementation of the 65nm dual-core 64b Merom processor, IEEE International Solid-State Circuits Conference Digest of Technical Papers, San Francisco, Cal., February 11-15, 2007, pp. 106-107; Marc Tremblay and Shailender Chaudhry, 2008, A third-generation 65nm 16-core 32-thread plus 32-count-thread CMT SPARC processor, IEEE International Solid-State Circuits Conference Digest of Technical Papers, San Francisco, Cal., February 3-7, 2008, p. 82-83; Larry Seiler, Doug Carmean, Eric Sprangle, Tom Forsyth, Michael Abrash, Pradeep Dubey, Stephen Junkins, Adam Lake, Jeremy Sugerman, Robert Cavin, Roger Espasa, Ed Grochowski, Toni Juan, and Pat Hanrahan, 2008, "Larrabee: A many-core x86 architecture for visual computing, ACM Transactions on Graphics 27(3): 1-15; Doug Carmean, 2008, Larrabee: A many-core x86 architecture for visual computing, Hot Chips 20: A Symposium on High Performance Chips, Stanford, Cal., August 24-26, 2008.

technology but consumes 23 times more power[23] and spends about 4 times more energy per instruction. That is another way of showing why single-processor power requirements increased because of circuit and architectural changes to improve performance. In achieving higher performance, the designs' energy efficiency was worsening: performance scaled because of technology scaling and growing power budgets.

CMPs provide an alternative approach: using less aggressive processor-core design to reduce energy dissipation per instruction and at the same time using multiple-processor cores to scale overall chip performance. That approach allows one to use the growing number of transistors per chip to scale performance while staying within the limit of air-cooling. It increases chip parallelism, but only a specific type of coarse-grain program parallelism can exploit this type of parallelism.

Switching to chip multiprocessors reduces the effect of wire delays (the length of time it takes a signal—output from a gate—to travel along a given length of wire), which is growing relative to the gate delay (the length of time it takes to translate input to a logic gate to be transformed into output from that gate).[24,25] Each processor in a CMP is small relative to the total chip area, and wires within a processor are short compared with the overall chip size. Interprocessor communication still requires long wires, but the latency of interprocessor communication is less critical for performance in a CMP system than is the latency between units within a single processor. In addition, the long wires can be pipelined and thus do not affect the clock-cycle time and performance of individual processors in a CMP.

Chip multiprocessors are a promising approach to scaling, but they face challenges as well; problems with modern scaling are described in the next section. Moreover, they cannot be programmed with the techniques that have proved successful for single processors; to achieve the potential performance of CMP, new software approaches and ultimately parallel applications must be developed. This will be discussed in the next chapter.

---

[23]Ed Grochowski, Ronny Ronen, John Shen, and Hong Wang., 2004, Best of both latency and throughput, Proceedings of the IEEE International Conference on Computer Design, San Jose, Cal., October 11-13, 2004, pp. 236-243.

[24]Mark Horowitz and William Dally, 2004, How scaling will change processor architecture, IEEE International Solid States Circuits Conference Digest of Technical Papers, San Francisco, Cal., February 15-19, 2004, pp. 132-133

[25]Kunle Olukotun, Basem A. Nayfeh, Lance Hammond, Ken Wilson, and Kunyung Chang, 1996, The case for a single-chip multiprocessor, Proceedings of 7th International Conference on Architectural Support for Programming Languages and Operating Systems, Cambridge, Mass., October 1-5, 1996, pp. 2-11.

PROBLEMS IN SCALING NANOMETER DEVICES

If voltages could continue to be scaled with feature size (following classic Dennard scaling), CMP performance could continue to be scaled with technology. However, early in this decade scaling ran into some fundamental limits that make it impossible to continue along that path,[26] and the improvements in both performance and power achieved with technology scaling have slowed from their historical rates. The net result is that even CMPs will run into power limitations. To understand those issues and their ramifications, we need to revisit technology scaling and look at one aspect of transistor performance that we ignored before: leakage current.

As described earlier, CMOS circuits have the important property that they dissipate energy only when a node changes value. Consider the simple but representative CMOS logic circuits in Figure 3.4. One type of CMOS device, a pMOS transistor, is connected to the power supply (V_{supply}). When its input is low (V_{gnd}), it turns on, connects V_{supply} to the output, and drives the output high to V_{supply}. When the input to the pMOS device is high (V_{supply}), it disconnects the output from V_{supply}. The other type of CMOS device, an nMOS transistor, has the complementary behavior: when its input is high (V_{supply}), it connects the output to V_{gnd}; when its input is low (V_{gnd}), it disconnects the output from V_{gnd}. Because of the construction of the CMOS logic, the pMOS and nMOS transistors are never driving the output at the same time. Hence, the only current that flows through the gate is that needed to charge or discharge the capacitances associated with the gate, so the energy consumed is mostly the energy needed to change the voltage on a capacitor with transistors, which is C_{load} multiplied by V_{supply}^2. For that analysis to hold, it is important that the off transistors not conduct any current in the off state: that is, they should have low leakage.

However, the voltage scaling that the industry has been following has indirectly been increasing leakage current. Transistors operate by changing the height of an energy barrier to modulate the number of carriers that can flow across them. One might expect a fairly sharp current transition, so that when the barrier is higher than the energy of the carriers, there is no current, and when it is lowered, the carriers can "spill" over and flow across the transistor. The actual situation is more complex. The basic reason is related to thermodynamics. At any finite temperature, although

---

[26]Sam Naffziger reviews the V_{dd} limitations and describes various approaches (circuit, architecture, and so on) to future processor design given the voltage scaling limitations in the article High-performance processors in a power-limited world, Proceedings of the IEEE Symposium on VLSI Circuits, Honolulu, Hawaii, June 15-17, 2006, pp. 93-97, available online at http://ewh.ieee.org/r5/denver/sscs/Presentations/2006_11_Naffziger_paper.pdf.

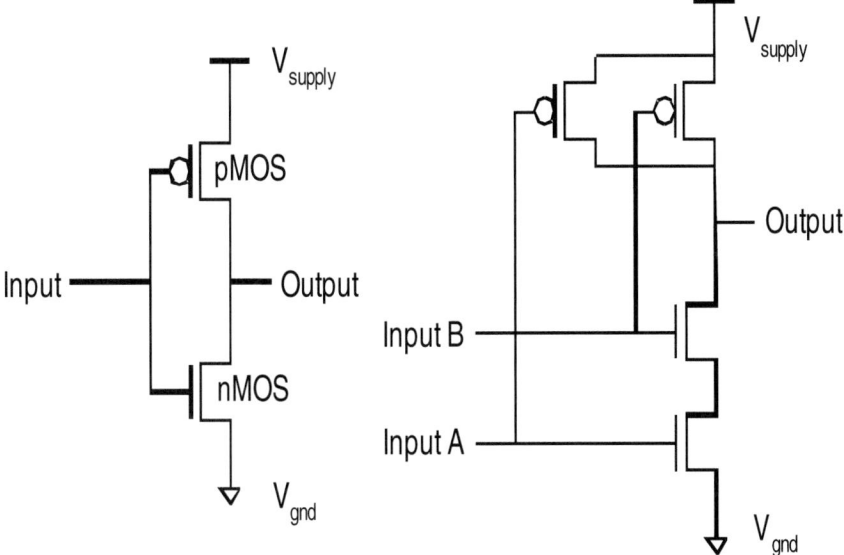

FIGURE 3.4 Representative CMOS logic circuits.

there is a well-defined average energy for the carriers, the energy of each individual carrier follows a probability distribution. The probability of having an energy higher than the average falls off exponentially, with a characteristic scale factor that is proportional to the temperature of the transistors measured measured in kelvins. The hotter the device, the wider the range of energies that the carriers can have.

That energy distribution is critical in the building of transistors. Even with an energy barrier that is higher than the average energy of the carriers, some carriers will flow over the barrier and through the transistor; the transistor will continue to conduct some current when we would like it to be off. The energy scale is kT, where k is the Boltzmann constant and T is the temperature in kelvins. We can convert it into voltage by dividing energy by the charge on the particle, an electron in this case: q = 1.6 × 10⁻¹⁹ coulombs. kT/q is around 26 mV at room temperature. Thus, the current through an off transistor drops exponentially with the height of the energy barrier, falling by slightly less than a factor of 3 for each 26-mV increase in the barrier height. The height of the barrier is normally called the threshold voltage (V_{th}) of the transistor, and the leakage current can be written as

$$I_{ds} = I_o e^{\frac{q(V_{gs} - V_{th})}{\alpha kT}},$$

where I_o is a constant around 1 µA (microampere) per micrometer of transistor width at room temperature, V_{gs} is the voltage applied to the control gate of the transistor, and α is a number greater than 1 (generally around 1.3) that represents how effectively the gate voltage changes the energy barrier. From the equation, it is easy to see that the amount of leakage current through an off transistor ($V_{gs} = 0$) depends heavily on the transistor's threshold voltage. The leakage current increases by about a factor of 10 each time the threshold voltage drops by another 100 mV.

Historically, V_{th}s were around 800 mV, so the residual transistor leakage currents were so small that they did not matter. Starting from high V_{th} values, it was possible to scale V_{th}, V_{supply}, and L together. While leakage current grew exponentially with shrinking V_{th}, the contribution of subthreshold leakage to the overall power was negligible as long as V_{th} values were still relatively large. But ultimately by the 90-nm node, the leakage grew to a point where it started to affect overall chip power.[27] At that point, V_{th} and V_{supply} scaling slowed dramatically.

One approach to reduce leakage current is to reduce temperature, inasmuch as this makes the exponential slope steeper. That is possible and has been tried on occasion, but it runs into two problems. The first is that one needs to consider the power and cost of providing a low-temperature environment, which usually dwarf the gains provided by the system; this is especially true for small or middle-size systems that operate in an office or home environment. The second is related to testing, repair, thermal cycling, and reliability of the systems. For those reasons, we will not consider this option further in the present report. However, for sufficiently large computing centers, it may prove advantageous to use liquid cooling or other chilling approaches where the energy costs of operating the semiconductor hardware in a low-temperature environment do not outweigh the performance gains, and hence energy savings, that are possible in such an environment.

V_{th} stopped scaling because of increasing leakage currents, and V_{supply} scaling slowed to preserve transistor speed with a constant ($V_{supply} - V_{th}$). Once leakage becomes important, an interesting optimization between V_{supply} and V_{th} is possible. Increasing V_{th} decreases leakage current but also makes the gates slower because the number of carriers that can flow through a transistor is roughly proportional to the decreasing ($V_{supply} - V_{th}$). One can recover the lost speed by increasing V_{supply}, but this also increases the power consumed to switch the gate dynamically. For a given gate delay, the lowest-power solution is one in which the marginal energy cost of increasing V_{dd} is exactly balanced by the marginal energy savings

---

[27]Edward J. Nowak, 2002, Maintaining the benefits of CMOS scaling when scaling bogs down, IBM Journal of Research and Development 46(2/3): 169-180.

of increasing V_{th}. The balance occurs when the static leakage power is roughly 30 percent of the dynamic power dissipation.

This leakage-constrained scaling began at roughly the 130-nm technology node, and today both V_{supply} and V_{th} scaling have dramatically slowed; this has also changed how gate energy and speed scale with technology. The energy required to switch a gate is C multiplied by V_{supply}^2, which scales only as $1/\kappa$ if V_{supply} is not scaling. That means that technology scaling reduces the power by κ only if the scaled circuit is run at the same frequency. That is, if gate speed continued to increase, half the die (the size of the scaled circuit) would dissipate the same power as the full die in the previous generation and would operate κ times, that is 1.4 times, faster, much less than the three-fold performance increase we have come to expect. Clearly, that is not optimal, so many designers are scaling V_{dd} slightly to increase the energy savings. That works but lowers the gate speeds, so some parallelism is needed just to recover from the slowing single-thread performance. The poor scaling will eventually limit the performance of CMPs.

Combining the lessons of the last several sections of this chapter, the committee concluded that neither CMOS nor chip multiprocessors can overcome the power limits facing modern computer systems. That leads to another core conclusion of this report. Basic laws of physics and constraints on chip design mean that the growth in the performance of computer systems will become limited by their power and thermal requirements within the next decade. Optimists might hope that new technologies and new research could overcome that limitation and allow hardware to continue to drive future performance scaling akin to what we have seen with single-thread performance, but there are reasons for caution, as described in the next section.

Finding: The growth in the performance of computing systems—even if they are multiple-processor parallel systems—will become limited by power consumption within a decade.

ADVANCED TECHNOLOGY OPTIONS

If CMOS scaling, even in chip-multiprocessor designs, is reaching limits, it is natural to ask whether other technology options might get around the limits and eventually overtake CMOS, as CMOS did to nMOS and bipolar circuits in the 1980s. The answer to the question is mixed.[28] It

---

[28]Mark Bohr, Intel senior fellow, gave a plenary talk at ISSCC 2009 on scaling in an SOC world in which he argues that "our challenge . . . is to recognize the coming revolutionary

is clear that new technologies and techniques will be created and applied to scaled technologies, but these major advances—such as high-k gate dielectrics, low-K interconnect dielectrics, and strained silicon—will probably be used to continue technology scaling in general and not create a disruptive change in the technology. Recent press reports make it clear, for example, that Intel expects to be using silicon supplemented with other materials in future generations of chips.[29]

A recent study compared estimated gate speed and energy of transistors built with exotic materials that should have very high performance.[30] Although the results were positive, the maximum improvement at the same power was modest, around a factor of 2 for the best technology. Those results should not be surprising. The fundamental problem is that V_{th} does not scale, so it is hard to scale the supply voltage. The limitation on V_{th} is set by leakage of carriers over an energy barrier, so *any device that modulates current by changing an energy barrier should have similar limitations*. All the devices used in the study cited above used the same current-control method, as do transistors made from nanotubes, nanowires, graphene, and so on. To get around that limitation, one needs to change "the game" and build devices that work by different principles. A few options are being pursued, but each has serious issues that would need to be overcome before they could become practical.

One alternative approach is to stop using an energy barrier to control current flow and instead use quantum mechanical tunneling. That approach eliminates the problem with the energy tails by using carriers that are constrained by the energy bands in the silicon, which have fixed levels. Because there is no energy tail, they can have, in theory, a steep turnon characteristic. Many researchers are trying to create a useful device of this type, but there are a number of challenges. The first is to create a large enough current ratio in a small enough voltage range. The tunneling current will turn on rapidly, but its increase with voltage is not that rapid. Because a current ratio of around 10,000 is required, we need a device that can transition through this current range in a small voltage (<400 mV).

---

changes and opportunities and to prepare to utilize them (Mark Bohr, 2009, The new era of scaling in an SOC world, IEEE International Solid-State Circuits Conference, San Francisco, Cal., February 9, 2009, available online at http://download.intel.com/technology/architecture-silicon/ISSCC_09_plenary_paper_Bohr.pdf).

[29]Intel CEO Paul Ottelini was said to have declared that silicon was in its last decade as the base material of the CPU (David Flynn, 2009, Intel looks beyond silicon for processors past 2017, Apcmag.com, October 29, 2009, available online at http://apcmag.com/intel-looks-beyond-silicon-for-processors-past-2017.htm).

[30]Donghyun Kim, Tejas Krishnamohan1, and Krishna C. Saraswat, 2008, Performance evaluation of 15nm gate length double-gate n-MOSFETs with high mobility channels: III-V, Ge and Si, Electrochemical Society Transactions 16(11): 47-55.

Even if one can create a device with that current ratio, another problem arises. The speed of the gates depends on the transistor current. So not only do we need the current ratio, we also need devices that can supply roughly the same magnitude of current as CMOS transistors provide. Tunnel currents are often small, so best estimates indicate that tunnel FETs might be much slower (less current) than in CMOS transistors. Such slowness will make their adoption difficult.

Another group of researchers are trying to leverage the collective effort of many particles together to get around the voltage limits of CMOS. Recall that the operating voltage is set by the thermal energy (kT) divided by the charge on one electron, because that is the charged particle. If the charged particle had a charge of 2q, the voltage requirements would be half what it is today. That is the approach that nerve cells use to operate robustly at low voltages. The proteins in the voltage-activated ion channels have a charge that allows them to operate easily at 100 mV. Although some groups have been trying to create paired charge carriers, most are looking at other types of cooperative processes. The ion channels in nerves go though a physical change, so many groups are trying to build logic from nanorelays (nanomicroelectromechanical systems, or nano MEMS). Because of the large number of charges on the gate electrode and the positive feedback intrinsic in electrostatic devices, it is theoretically possible to have very low operating voltages; indeed, operation down to a couple of tenths of a volt seems possible. Even as researchers work to overcome that hurdle, there are a number of issues that need to be addressed. The most important is determining the minimum operating voltage that can reliably overcome contact sticking. It might not take much voltage to create a contact, but if the two surfaces that connect stick together (either because of molecular forces or because of microwelding from the current flow), larger voltages will be needed to break the contact. Devices will have large numbers of these structures, so the voltage must be less than CMOS operating voltages at similar performance. The second issue is performance and reliability. This device depends on a mechanically moving structure, so the delay will probably be larger than that of CMOS (around 1 nanosecond), and it will probably be an additional challenge to build structures that can move for billions of cycles without failing.

There is also promising research in the use of electron-spin-based devices (spintronics) in contrast with the charge-based devices (electronics) in use today. Spin-based devices—and even pseudospin devices, such as the BiSFET[31]—have the potential to greatly reduce the power dissi-

---

[31] Sanjay K. Banerjee, Leonard F. Register, Emanuel Tutuc, Dharmendar Reddy, and Allan H. MacDonald, 2009, Bilayer pseudospin field-effect transistor (BiSFET): A proposed new logic device, IEEE Electron Device Letters 30(2): 158-160.

pated in performing basic logic functions. However, large fundamental and practical problems remain to be solved before spintronic systems can become practical.[32] Those or other approaches (such as using the correlation of particles in ferro materials[33]) might yield a breakthrough. However, given the complexity of today's chips, with billions of working transistors, it is likely to take at least a decade to introduce any new technology into volume manufacturing. Thus, although we should continue to invest in technology research, we cannot count on it to save the day. It is unlikely to change the situation in the next decade.

Recommendation: Invest in research and development to make computer systems more power-efficient at all levels of the system, including software, application-specific approaches, and alternative devices. R&D should be aimed at making logic gates more power-efficient. Such efforts should address alternative physical devices beyond incremental improvements in today's CMOS circuits.

APPLICATION-SPECIFIC INTEGRATED CIRCUITS

Although the shift toward chip multiprocessors will allow industry to continue to scale the performance of CMPs based on general-purpose processor cores for some time, general-purpose chip multiprocessors will reach their own limit. As discussed earlier, CMP designers can trade off single-thread performance of individual processors against lower energy dissipation per instruction, thus allowing more instructions by multiple processors while the same amount of energy is dissipated by the chip. However, that is possible only within some range of energy performance. Beyond some limit, lowering energy per instruction by processor simplification can lead to overall CMP performance degradation because processor performance starts to decrease faster than energy per instruction. That range is likely to be a factor of about 10, that is, energy per instruction cannot be reduced by more than a factor of 10 compared with the highest-performance single-processor chip, such as the Intel Pentium 4 or the Intel Itanium.[34]

When such limits are reached, we will need to create other approaches

---

[32]In their article, cited in the preceding footnote, Banerjee et al. look at a promising technology that still faces many challenges.

[33]See, for instance, the research of Sayeef Salahuddin at the University of California, Berkeley.

[34]The real gain might be even smaller because with an increase in the number of processors on the chip, more energy will be dissipated by the memory system and interconnect, or the performance of many parallel applications will scale less than linearly with the number of processors.

to create an energy-efficient computation unit. On the basis of the historical data, the answer seems clear: we will need to create more application-optimized processing units. It is well known that tuning the hardware and software toward a specific application or set of applications allows a more energy-efficient solution. That work started with the digital watch many decades ago and continues today. Figure 3.5 shows data for general-purpose processors, digital-signal processors, and application-specific integrated circuits (ASICs) from publications presented at the International Solid-State Circuits Conference. The data are somewhat dated, but all chips were designed for similar 0.18- to 0.25-µm CMOS technology, and one can see that the ASIC designs are roughly 3 orders of magnitude more energy-efficient than the general-purpose processors.

The main reason for such a difference is a combination of algorithm and hardware tuning and the ability to reduce the use of large memory structures as general interconnects: instead of a value's being stored in a register or memory, it is consumed by the next function unit. Doing only what needs to be done saves both energy and area (see Figure 3.6).

More recently, researchers at Lawrence Berkeley National Laboratory, interested in building peta-scale supercomputers for kilometer-scale climate modeling, argued that designing a specialized supercomputer based on highly efficient customizable embedded processors can be attractive in terms of energy cost.[35] For example, they estimated that a peta-scale climate supercomputer built with custom chips would consume 2.5 MW of electric power whereas a computer with the same level of performance but built with general-purpose AMD processors would require 179 MW.

The current design trend, however, is away from building customized solutions; increasing design complexity has caused the *nonrecurring engineering* costs for designing these chips to grow rapidly. Typical ASIC design requires $20-50 million, which limits the range of market segments to very few with volumes high enough to justify the initial engineering investment. Thus, if we do need to create more application-optimized computing systems, we will need to create a new approach to design that will allow a small team to create an application-specific chip at reasonable cost. That leads to this chapter's overarching recommendation. Efforts are needed along multiple paths to deal with the power limitations that modern scaling and computer-chip designs are encountering.

---

[35]Michael Wehner, Leonid Oliker, and John Shalf, 2008, Towards ultra-high resolution models of climate and weather, International Journal of High Performance Computing Applications 22(2): 149-165.

Chip #	Year	Paper	Description
1	1997	10.3	µP - S/390
2	2000	5.2	µP – PPC (SOI)
3	1999	5.2	µP - G5
4	2000	5.6	µP - G6
Microprocessors			
5	2000	5.1	µP - Alpha
6	1998	15.4	µP - P6
7	1998	18.4	µP - Alpha
8	1999	5.6	µP – PPC
9	1998	18.6	DSP - StrongArm
DSP's			
10	2000	4.2	DSP – Comm

Chip #	Year	Paper	Description
11	1998	18.1	DSP -Graphics
12	1998	18.2	DSP - Multimedia
DSPs			
13	2000	14.6	DSP – Multimedia
14	2002	22.1	DSP – Mpeg Decoder
15	1998	18.3	DSP - Multimedia
16	2001	21.2	Encryption Processor
17	2000	14.5	Hearing Aid Processor
Dedicated			
18	2000	4.7	FIR for Disk Read Head
19	1998	2.1	MPEG Encoder
20	2002	7.2	802.11a Baseband

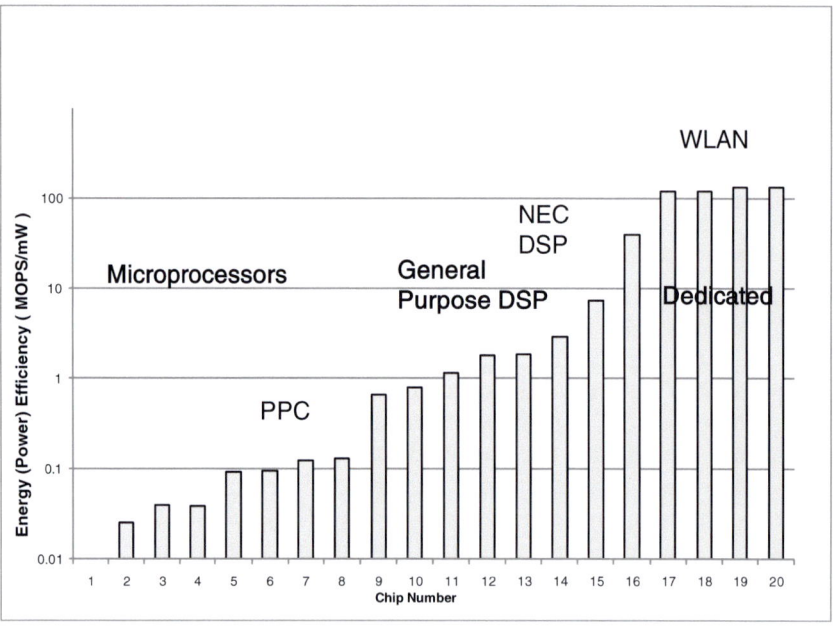

FIGURE 3.5 Energy efficiency comparison of CPUs, DSPs, and ASICs. SOURCE: Robert Brodersen of the University of California, Berkeley, and Teresa Meng of Stanford University. Data published at International Solid-State Circuits Conference (0.18- to 0.25-µm).

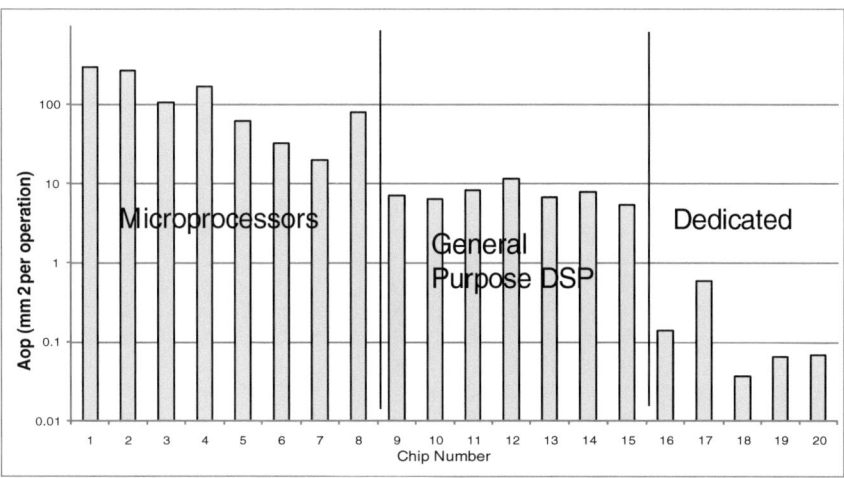

FIGURE 3.6 Area efficiency comparison of CPUs, DSPs, and ASICs. SOURCE: Robert Brodersen of the University of California at Berkeley and Teresa Meng of Stanford University.

Recommendation: Invest in research and development of parallel architectures driven by applications, including enhancements of chip multiprocessor systems and conventional data-parallel architectures, cost-effective designs for application-specific architectures, and support for radically different approaches.

BIBLIOGRAPHY

Broderson, Robert. "Interview: A Conversation with Teresa Meng." *ACM Queue* 2(6): 14-21, 2004.
Chabini, Noureddine, Ismaïl Chabini, El Mostapha Aboulhamid, and Yvon Savaria. "Methods for Minimizing Dynamic Power Consumption in Synchronous Designs with Multiple Supply Voltages." In *IEEE Transactions on Computer-Aided Design of Integrated Circuits and Systems* 22(3): 346-351, 2003.
Copeland, Jack. *Colossus: The Secrets of Bletchley Park's Code-Breaking Computers.* New York: Oxford University Press, 2006.
International Technology Roadmap for Semiconductors (ITRS). "System Drivers." *ITRS 2007 Edition.* Available online at http://www.itrs.net/Links/2007ITRS/Home2007.htm.
Kannan, Hari, Fei Guo, Li Zhao, Ramesh Illikkal, Ravi Iyer, Don Newell, Yan Solihin, and Christos Kozyrakis. "From Chaos to QoS: Case Studies in CMP Resource Management." At the 2nd Workshop on Design, Architecture, and Simulation of Chip-Multiprocessors (dasCMP). Orlando, Fla., December 10, 2006.
Khailany, Brucek, William J. Dally, Scott Rixner, Ujval J. Kapasi, Peter Mattson, Jinyung Namkoong, John D. Owens, Brian Towles, and Andrew Chang. "Imagine: Media Processing with Streams." *IEEE Micro* 21(2): 35-46, 2001.

Knight, Tom. 1986."An Architecture for Mostly Functional Languages." In *Proceedings of ACM Conference on LISP and Functional Programming.* Cambridge, Mass., August 4-6, 1986, pp. 105-112.

Poonacha Kongetira, Kathirgamar Aingaran, and Kunle Olukotun. "Niagara: A 32-way Multithreaded Sparc Processor." *IEEE Micro* 25(2):21-29 2005, 2005.

Lee, Edward A. "The Problem with Threads." *IEEE Computer* 39(5): 33-42, 2006.

Lee, Walter, Rajeev Barua, Matthew Frank, Devabhaktuni Srikrishna, Jonathan Babb, Vivek Sarkar, and Saman Amarasinghe. "Space-time Scheduling of Instruction-level Parallelism on a Raw Machine." In *Proceedings of the Eighth International Conference on Architectural Support for Programming Language and Operating Systems.* San Jose, Cal., October 3-7, 1998, pp. 46-57.

Lomet, David B., "Process Structuring, Synchronization, and Recovery Using Atomic Actions," In *Proceedings of the ACM Conference on Language Design for Reliable Software.* Raleigh, N.C., March 28-30, 1977, pp. 128-137.

Marković, Dejan, Borivoje Nikolić, and Robert W. Brodersen. "Power and Area Minimization for Multidimensional Signal Processing." *IEEE Journal of Solid-State Circuits* 42(4): 922-934, 2007.

Nowak, Edward J. "Maintaining the Benefits of CMOS Scaling When Scaling Bogs Down." *IBM Journal of Research and Development* 46(2/3):169-180, 2002.

Rixner, Scott, William J. Dally, Ujval J. Kapasi, Brucek Khailany, Abelardo López-Lagunas, Peter R. Mattson, and John D. Owens. "A Bandwidth-Efficient Architecture for Media Processing." In *Proceedings of the International Symposium on Microarchitecture.* Dallas, Tex.: November 30-December 2, 1998, pp. 3-13, 1998.

Rusu, Stefan, Simon Tam, Harry Muljono, David Ayers, and Jonathan Chang. "A Dual-core Multi-threaded Xeon Processor with 16MB L3 Cache." In *IEEE International Solid-State Circuits Conference Digest of Technical Papers.* San Francisco, Cal., February 6-9, 2006, pp. 315-324.

Sandararajan, Vijay, and Keshab Parhi. "Synthesis of Low Power CMOS VLSI Circuits Using Dual Supply Voltages." In *Proceedings of the 35*th *Design Automation Conference.* New Orleans, La.., June 21-25, 1999, pp. 72-75.

Sutter, Herb, and James Larus. "Software and the Concurrency Revolution." *ACM Queue* 3(7): 54-62, 2005.

Taur, Yuan, and Tak H. Ning, *Fundamentals of Modern VLSI Devices*, Ninth Edition, New York: Cambridge University Press, 2006.

Thies, Bill, Michal Karczmarek, and Saman Amarasinghe. "StreamIt: A Language for Streaming Applications." In *Proceedings of the International Conference on Compiler Construction.* Grenoble, France, April 8-12, 2002, pp. 179-196.

Wehner, Michael, Leonid Oliker, and John Shalf. "Towards Ultra-High Resolution Models of Climate and Weather." *International Journal of High Performance Computing Application* 22(2): 149-165, 2008.

Zhao, Li, Ravi Iyer, Ramesh Illikkal, Jaideep Moses, Srihari Makineni, and Don Newell. "CacheScouts: Fine-Grain Monitoring of Shared Caches in CMP Platforms." In *Proceedings of the 16th International Conference on Parallel Architecture and Compilation Techniques.* Brasov, Romania, September 15-19, 2007, pp. 339-352.

4

The End of Programming as We Know It

Future growth in computing performance will have to come from software parallelism that can exploit hardware parallelism. Programs will need to be expressed by dividing work into multiple computations that execute on separate processors and that communicate infrequently or, better yet, not at all. This chapter first explains how current software reaped the benefits of Moore's law and how much of the resulting software is not well suited to parallelism. It then explores the challenges of programming parallel systems. The committee explores examples of software and programming parallelism successes and possibilities for leveraging these successes, as well as examples of limitations of parallelism and challenges to programming parallel systems. The sudden shift from single-core to multiple-core processor chips requires a dramatic change in programming, but software developers are also challenged by the continuously widening gap between memory system and processor performance. That gap is often referred to as the "memory wall," but it reflects a continuous rather than discrete shift in the balance between the costs of computational and memory operations and adds to the difficulty of obtaining high performance. To optimize for locality, software must be written to minimize both communication between processors and data transfers between processors and memory.

MOORE'S BOUNTY: SOFTWARE ABSTRACTION

Moore's bounty is a portable sequential-programming model.[1] Programmers did not have to rewrite their software—software just ran faster on the next generation of hardware. Programmers therefore designed and built new software applications that executed correctly on current hardware but were often too compute-intensive to be useful (that is, it took too long to get a useful answer in many cases), anticipating the next generation of faster hardware. The software pressure built demand for next-generation hardware. The sequential-programming model evolved in that ecosystem as well. To build innovative, more capable sophisticated software, software developers turned increasingly to higher-level sequential-programming languages and higher levels of abstraction (that is, the reuse of libraries and component software for common tasks). Moore's law helped to drive the progression in sequential-language abstractions because increasing processor speed hid their costs.

For example, early sequential computers were programmed in assembly language. Assembly-language statements have a one-to-one mapping to the instructions that the computer executes. In 1957, Backus and his colleagues at IBM recognized that assembly-language programming was arduous, and they introduced the first implementation of a high-level sequential scientific computing language, called Fortran, for the IBM 704 computer.[2] Instead of writing assembly language, programmers wrote in Fortran, and then a compiler translated Fortran into the computer's assembly language. The IBM team made the following claims for that approach:

1. Programs will contain fewer errors because writing, reading, and understanding Fortran is easier than performing the same tasks in assembly language.
2. It will take less time to write a correct program in Fortran because each Fortran statement is an abstraction of many assembly instructions.
3. The performance of the program will be comparable with that of assembly language given good compiler technology.

---

[1] Jim Larus makes this argument in an article in 2009, Spending Moore's dividend, Communications of the ACM 52(5): 62-69.

[2] J.W. Backus, R.I. Beeber, S. Best, R. Goldberg, L.M. Haibt, H.L. Herrick, R.A. Nelson, D. Sayre, P.B. Sheridan, H. Stern, Ziller, R.A. Hughes, and R. Nutt, 1957, The Fortran automatic coding system, Proceedings of the Western Joint Computer Conference, Los Angeles, Cal., pp. 187-198, available online at http://archive.computerhistory.org/resources/text/Fortran/102663113.05.01.acc.pdf.

The claimed benefits of high-level languages are now widely accepted. In fact, as computers got faster, modern programming languages added more and more abstractions. For example, modern languages—such as Java, C#, Ruby, Python, F#, PHP, and Javascript—provide such features as automatic memory management, object orientation, static typing, dynamic typing, and referential transparency, all of which ease the programming task. They do that often at a performance cost, but companies chose these languages to improve the correctness and functionality of their software, which they valued more than performance mainly because the progress of Moore's law hid the costs of abstraction. Although higher levels of abstraction often result in performance penalties, the initial transition away from hand-coded assembly language came with performance gains, in that compilers are better at managing the complexity of low-level code generation, such as register allocation and instruction scheduling, better than most programmers.

That pattern of increases in processor performance coupling with increases in the use of programming-language abstractions has played out repeatedly. The above discussion describes the coupling for general-purpose computing devices, and it is at various stages in other hardware devices, such as graphics hardware, cell phones, personal digital assistants, and other embedded devices.

High-level programming languages have made it easier to create capable, large sophisticated sequential programs. During the height of the synergy between software and increasing single-processor speeds in 1997, Nathan Myhrvold, former chief technology officer for Microsoft, postulated four "laws of software":[3]

1. Software is a gas. Software always expands to fit whatever container it is stored in.
2. Software grows until it becomes limited by Moore's law. The growth of software is initially rapid, like gas expanding, but is inevitably limited by the rate of increase in hardware speed.
3. Software growth makes Moore's law possible. People buy new hardware because the software requires it.
4. Software is limited only by human ambition and expectation. We will always find new algorithms, new applications, and new users.

---

[3]These laws were described in a 1997 presentation that the Association for Computing Machinery hosted on the next 50 years of computing (Nathan P. Myhrvold, 1997, The next fifty years of software, Presentation, available at http://research.microsoft.com/en-us/um/siliconvalley/events/acm97/nmNoVid.ppt).

Myhrvold's analysis explains both the expansion of existing applications and the explosive growth in innovative applications. Some of the code expansion can be attributed to a lack of attention to performance and memory use: it is often easier to leave old and inefficient code in a system than to optimize it and clean it up. But the growth in performance also enabled the addition of new features into existing software systems and new paradigms for computing. For example, Vincent Maraia reports that in 1993, the Windows NT 3.1 operating system (OS) consisted of 4-5 million lines of code and by 2003, the Windows Server OS had 50 million lines, 10 times as many.[4] Similarly, from 2000 to 2007, the Debian 2.2 Linux OS grew from about 59 to 283 million lines in version 4.0, about 5 times as many.[5] Those operating systems added capabilities, such as better reliability, without slowing down the existing features, and users experienced faster operating system startup time and improvement in overall performance. Furthermore, the improvements described by Moore's law enabled new applications in such domains as science, entertainment, business, and communication. Thus, the key driver in the virtuous cycle of exploiting Moore's law is that applications benefited from processor performance improvements without those applications having to be adapted to changes in hardware. *Programs ran faster on successive generations of hardware, allowing new features to be added without slowing the application performance.*

The problem is that much of the innovative software is sequential and is designed to execute on only one processor, whereas the previous chapters explained why all future computers will contain multiple processors. Thus, current programs will not run faster on successive generations of hardware.[6] The shift in the hardware industry has broken the performance-portability connection in the virtuous cycle—sequential programs will not benefit from increases in processor performance that stem from the use of multiple processors. There were and are many problems—for example, in search, Web applications, graphics, and scientific computing—that require much more processing capability than a single processor pro-

---

[4]Vincent Maraia, The Build Master: Microsoft's Software Configuration Management Best Practices, Addison-Wesley Professional, 2005.
[5]Debian Web site, Wikipedia.com.http://en.wikipedia.org/wiki/Debian.
[6]Successive generations of hardware processors will not continue to increase in performance as they have in the past; this may be an incentive for programmers to develop tools and methods to optimize and extract the most performance possible from sequential programs. In other words, working to eliminate the inefficiencies in software may yield impressive gains given that past progress in hardware performance encouraged work on new functions rather than optimizing existing functions. However, optimizing deployed software for efficiency will ultimately reach a point of diminishing returns and is not a long-term alternative to moving to parallel systems for more performance.

vides. The developers of the applications and programming systems have made much progress in providing appropriate abstractions (discussed in detail below) but not enough in that most developers and programming systems currently use the sequential model. Conventional sequential programs and programming systems are ill equipped to support parallel programming because they lack abstractions to deal with the problems of extracting parallelism, synchronizing computations, managing locality, and balancing load. In the future, however, *all* software must be able to exploit multiple processors to enter into a new virtuous cycle with successive generations of parallel hardware that expands software capabilities and generates new applications.[7]

Finding: There is no known alternative to parallel systems for sustaining growth in computing performance; however, no compelling programming paradigms for general parallel systems have yet emerged.

To develop parallel applications, future developers must invent new parallel algorithms and build new parallel applications. The applications will require new parallel-programming languages, abstractions, compilers, debuggers, execution environments, operating systems, and hardware virtualization systems. We refer to those tools collectively as a *programming system*. Future programming systems will need to take advantage of all those features to build applications whose performance will be able to improve on successive generations of parallel hardware that increase their capabilities by increasing the number of processors. In contrast, what we have today are conventional sequential-programming systems based on two abstractions that are fundamentally at odds with parallelism and locality. First, they tie the ordering of statements in a program to a serial execution order of the statements. Any form of parallelism violates that model unless it is unobservable. Second, conventional programs are written on the assumption of a flat, uniform-cost global memory system. Coordinating locality (minimizing the number of expensive main memory references) is at odds with the flat model of memory that does not distinguish between fast and slow memory (for example, on and off chip). Parallelism and locality are also often in conflict in that

---

[7]Indeed, the compiler community is bracing for the challenges ahead. On p. 62 of their book, Mary Hall et al. observe that "exploiting large-scale parallel hardware will be essential for improving an application's performance or its capabilities in terms of execution speed and power consumption. The challenge for compiler research is how to enable the exploitation of the power [that is, performance, not thermals or energy] of the target machine, including its parallelism, without undue programmer effort." (Mary Hall, David Padua, and Keshav Pingali, 2009, Compiler research: The next 50 years, Communications of the AC 52(2): 60-67.

locality will encourage designs that put all data close to a single processor to avoid expensive remote references, whereas performing computations in parallel requires spreading data among processors.

SOFTWARE IMPLICATIONS OF PARALLELISM

There are five main challenges to increasing performance and energy efficiency through parallelism:

- Finding independent operations.
- Communicating between operations.
- Preserving locality between operations.
- Synchronizing operations.
- Balancing the load represented by the operations among the system resources.

The first challenge in making an application parallel is to design a parallel algorithm to solve the problem at hand that provides enough independent operations to keep the available parallel resources busy. Some demanding problems have large amounts of *data parallelism*—that is, a single operation can be performed for every data element of a set, and the operations are independent of one another (or can be made so via transformations). Some problems also have moderate amounts of *control* or *task parallelism* in which different operations can be performed in parallel on different data items. In both task and data parallelism, an operation may comprise a sequence of instructions. For some applications, the parallelism is limited by a sequence of dependent operations, and performance is limited not by throughput but by the latency along this critical path.[8]

The second challenge, communication, occurs when computations that execute in parallel are not entirely independent and must communicate. Some demanding problems cannot be divided into completely independent parallel tasks, but they can be divided into parallel tasks that communicate to find a solution cooperatively. For example, to search for a particular object in an image, one may divide the image into pieces that are searched by independent tasks. If the object crosses pieces, the tasks will need to communicate. The programming system can perform communication through inputs and outputs along dependences by reading and writing to shared data structures or by explicitly sending messages

---

[8]This limit on parallelism is often called Amdahl's law, after Gene Amdahl. For more on this law, see Box 2.4.

between parallel operations. Even in the implicit case, some data will need to transfer between processors to allow access to shared data.

Locality, the third challenge, reduces the costs associated with communication by placing two operations that access the same data near each other in space or in time. Scheduling operations nearby in space on the same processor avoids communication entirely, and placing them on nearby processors may reduce the distance that data need to travel. Scheduling operations nearby in time shortens the lifetime of data produced by one operation and consumed by another; this reduces the volume of live data and allows the data to be captured in small on-chip memories. Locality avoids the need for communication between processors, but is also critical for avoiding another form of communication: the movement of data between memory and processors.

The fourth challenge, synchronization, is also needed to provide cooperation between parallel computations. Some operations must be performed in a particular order to observe dependence. Other operations may be performed in an arbitrary order but must be grouped so that some sequences execute *atomically* (without interference from other sequences). Synchronization is used to serialize parts of an otherwise parallel execution, and there is often a tension between the performance gained from parallelism and the correctness ensured by synchronization. For example, *barrier* synchronization forces a set of parallel computations to wait until all of them reach the barrier. Locks are used to control access to shared data structures by allowing only one thread to hold a lock at a given time. *Unnecessary* synchronization may occur when an entire data structure is locked to manipulate one element or when a barrier is placed on every loop iteration even when the iterations are independent.

Finally, load balancing involves distributing operations evenly among processors. If the load becomes unbalanced because some processors have more work than others or take more time to perform their work, other processors will be idle at barrier synchronization or when program execution ends. The difficulty of load balancing depends on the characteristics of the application. Load balancing is trivial if all parallel computations have the same cost, more difficult if they have different costs that are known in advance, and even more difficult if the costs are not known until the tasks execute.

Locality's Increasing Importance

Effective parallel computation is tied to coordinating computations and data; that is, the system must collocate computations with their data. Data are stored in memory. Main-memory bandwidth, access energy, and latency have all scaled at a lower rate than the corresponding characteristics of processor chips for many years. In short, there is an ever-widening gap between processor and memory performance. On-chip cache memories are used to bridge the gap between processor and memory performance partially. However, even a cache with the best algorithm to predict the next operands needed by the processor does not have a success rate high enough to close the gap effectively. The advent of chip multiprocessors means that the bandwidth gap will probably continue to widen in that the aggregate rate of computation on a single chip will continue to outpace main-memory capacity and performance improvements. The gap between memory latency and computation is also a limitation in software performance, although this gap will not grow with multicore technology, inasmuch as clock rates are relatively constant. In addition to performance concerns, the movement of data between cores and between the processor and memory chips consumes a substantial fraction of a system's total power budget. Hence, to keep memory from severely limiting system power and performance, applications must have locality, and we must increase the amount of locality. In other words, the mapping of data and computation should minimize the distance that data must travel.

To see the importance of locality in future systems, it is instructive to examine the relative energy per operation for contemporary systems and how it is expected to scale with technology. In a contemporary 40-nm CMOS process, performing a 64-bit floating-point multiply-add (FMA) operation requires that the energy of the operation, E_{op}, be equal to 100 pJ. The energy consumed in moving data over 1 mm of wire, E_w, is 200 fJ/bit-mm, or 12.8 pJ/W-mm (for 64-bit words). Moving data off-chip takes energy, E_p, of 2 pJ/bit (128 pJ/W) or more. Supplying the four operands (three input and one output) of the FMA operation from even 1 mm away takes 51.2 pJ—half as much energy as doing the operation itself. Supplying the data globally on-chip—say, over a distance of 20 mm—takes about 1 nJ, an order of magnitude more energy than doing the operation. Moving data off-chip is comparably expensive. Thus, to avoid having the vast majority of all energy be spent in moving data, it is imperative that data be kept local.

Locality is inherently present in many algorithms, but the computation must be properly ordered to express locality. For dense matrix computations, ordering is usually expressed by *blocking* the algorithm. For example, consider multiplying two 10,000 × 10,000 matrices. Using the straightforward algorithm, it requires performing 2×10^{12} arithmetic operations. If we perform the operations in a random order, there is little locality, and 4×10^{12} memory references will be required to compute the result, so both arithmetic operations and data access grow with the cube of the matrix dimension. Even with a natural implementation based on three nested loops, data accesses will grow with the cube of the matrix dimension, because one of the matrices will be accessed in an order that allows little reuse of data in the cache. However, if we decompose the problem into smaller matrix multiplication problems, we can capture locality, reusing each word fetched from memory many times.

Suppose we have a memory capable of holding 256 kB (32 kW) 1 mm from our floating-point unit. The local memory is large enough to hold three 100 × 100 submatrices, one for each input operand and one for the partial result. We can perform a 100 × 100 matrix multiplication entirely out of the local memory, performing 2×10^6 operations with only 4×10^4 memory references—a ratio of 50 operations per reference. We can apply this blocking recursively. If there is aggregate on-chip memory of 32 MB (4 MW), we can hold three 1,000 × 1,000 submatrices at this level of the storage hierarchy. In a seminal paper by Hong and Kung, that idea was proved to be optimal for matrix multiplication in the sense that this kind of blocked algorithm moves the minimum amount of data possible between processor and memory system.[9] Other array computations, including convolutions and fast Fourier transformations, can be blocked in this manner—although with different computation-to-communication ratios—and there are theoretical results on the optimality of communication for several linear algebra problems for both parallel and serial machines.

The recursive nature of the blocked algorithm also led to the notion of "cache-oblivious" algorithms, in which the recursive subdivision produces successively smaller subproblems that eventually fit into a cache or other fast memory layer.[10] Whereas other blocked algorithms are implemented to match the size of a cache, the oblivious algorithms are optimized for locality without having specific constants, such as cache size, in their implementation. Locality optimizations for irregular codes, such as graph algorithms, can be much more difficult because the data structures are built with pointers or indexed structures that lead to random memory accesses. Even some of the graph algorithms have considerable locality that can be realized by partitioning the graph subgraphs that fit into a local memory and reorganizing the computations to operate on each subgraph with reuse before moving on to the next subgraph. There are many algorithms and software libraries for performing graph partition-

---

[9]See Hong Jia-Wei and H.T. Kung, 1981, I/O complexity: The red-blue pebble game, Proceedings of the Thirteenth Annual ACM Symposium on Theory of Computing, Milwaukee, Wis., May 11-13, 1981, pp. 326-333

[10]Matteo Frigo, Charles E. Leiserson, Harald Prokop, and Sridhar Ramachandran, 1999, Cache-oblivious algorithms, Proceedings of the 40th IEEE Symposium on Foundations of Computer Science, New York, N.Y., October 17-19, 1999, pp. 285-297.

ing that minimize edge cuts for locality but with equal subgraph sizes for load-balancing.[11]

A key challenge in exploiting locality is developing abstractions for locality that allow a programmer to express the locality in a program independent of any particular target machine. One promising approach, used by the Sequoia programming system,[12] is to present the programmer with an abstract memory hierarchy. The programmer views the machines as a tree of memories; the number of levels in the tree and the size of the memory at each level are unspecified. The programmer describes a decomposition method that subdivides the problem at one level into smaller problems at the next level and combines the partial solutions and a leaf method that solves the subproblem at the lowest level of the hierarchy. An autotuner then determines the number of times to apply the decomposition method and the appropriate data sizes at each level to map the program optimally onto a specific machine. The result is a programming approach that gives good locality with portability among diverse target machines.

Software Abstractions and Hardware Mechanisms Needed

Simplifying the task of parallel programming requires software abstractions that provide powerful mechanisms for synchronization, load balance, communication, and locality, as described above, while hiding the underlying details. Most current mechanisms for these operations are low-level and architecture-specific. The mechanisms must be carefully programmed to obtain good performance with a given parallel architecture, and the resulting programs are typically not performance-portable; that is, they do not exhibit better performance with a similar parallel architecture that has more processors. Successful software abstractions are needed to enable programmers to express the parallelism that is inherent in a program and the dependences between operations and to structure a program to enhance locality without being bogged down in low-level architectural details. Which abstractions make parallel programming convenient and result in performance-portable programs is an open research question. Successful abstractions will probably involve global address spaces, accessible ways to describe or invoke parallel operations over

---

[11]For one example of a graph-partitioning library, see George Karypis and Vipin Kumar, 1995, METIS: Unstructured Graph Partitioning and Sparse Matrix Ordering System, Technical report, Minneapolis, Minn.: University of Minnesota.

[12]See Kayvon Fatahalian, Timothy J. Knight, Mike Houston, Mattan Erez, Daniel Reiter Horn, Larkhoon Leem, Ji Young Park, Manman Ren, Alex Aiken, William J. Dally, and Pat Hanrahan, 2006, Sequoia: Programming the memory hierarchy, Proceedings of the ACM/IEEE Conference on Supercomputing, Tampa, Fla., November 11-17, 2006.

collections of data, and constructs for atomic operations. Abstractions may also involve abstract machine models that capture resource costs and locality while hiding details of particular machines. Abstractions for parallelism are typically encapsulated in a programming system and execution model.

At the same time, reasonable performance requires efficient underlying hardware mechanisms, particularly in cases that need fine-grained communication and synchronization. Some parallel machines require interactions between processors to occur by means of high-overhead message transfers or by passing data via shared memory locations. Such mechanisms are useful but can be cumbersome and restrict the granularity of parallelism that can be efficiently exploited. Resolving those details will require research, but successful mechanisms will enable low-overhead communication and synchronization and will facilitate migration of data and operations to balance load. There are several emerging directions in hardware to support parallel computations. It is too early to know which hardware architecture or architectures will prove most successful, but several trends are evident:

- Multiple processors sharing a memory. This direction was taken by chip multiprocessors and was the primary approach used by semiconductor companies once they could not continue to increase their single-processor products.
- Multiple computers interconnected via a high-speed communication network. When very large computation facilities are needed for research or business, it is impractical for all the processors to share a memory, and a high-speed interconnect is used to tie the hundreds or thousands of processors together in a single system. Data centers use this model.
- A single processor containing multiple execution units. In this architecture, a single processor, or instruction stream, controls an array of similar execution units. This is sometimes termed single-instruction stream multiple-data (SIMD) architecture.
- Array of specialized processors. This approach is effective for executing a specialized task, such as a graphic or video processing algorithm. Each individual processor and its interconnections can be tailored and simplied for the target application.
- Field-programmable gate arrays (FPGAs) used in some parallel computing systems. FPGAs with execution units embedded in their fabric can yield high performance because they exploit locality and program their on-chip interconnects to match the data flow of the application.

That list of parallel architectures is not exhaustive, and some systems will use a combination of them. We expect current versions of the architectures to evolve substantially to support the most promising programming systems, and we may see entirely new hardware architectures in support of not-yet-developed programming approaches to parallel computing.

An encouraging development is that programs of research in parallelism being initiated or revived in a few research universities. Some research efforts already under way are aimed at some of the challenges that this report outlines. For example, in 2008, the University of California, Berkeley, and the University of Illinois at Urbana–Champaign were awarded research grants from Microsoft and Intel to establish Universal Parallel Computing Research Centers. In 2009, Stanford University—with industrial funding from Sun, AMD, NVIDIA, and other companies—started the Pervasive Parallelism Laboratory. Those centers at leading research universities are a good beginning to address the broad and challenging research agenda that we outline below, but they are just a beginning. History shows that the development of technology similar to that needed for parallelism often takes a decade or more. The results of such research are needed now, so the research is starting a decade late. Moreover, there is no guarantee that there is an answer to the challenges. If there is not a good answer, we need to know that as soon as possible so that we can push innovation in some other direction in a timely way.

THE CHALLENGES OF PARALLELISM

Parallelism has long been recognized as promising to achieve greater computing performance. Research on parallel hardware architectures began in earnest in the 1960s.[13] Many ways of organizing computers have been investigated, including vector machines, SIMD machines, shared-memory multiprocessors, very-long-instruction-word machines, data-flow machines, distributed-memory machines, nonuniform-memory architectures, and multithreaded architectures. As described elsewhere in this report, single-processor performance has historically been making it difficult exponentially for companies promoting specialized parallel architectures to succeed. Over the years, however, ideas that have originated or been refined in the parallel-computer architecture research community have become standard features on PC processors, such as having SIMD instructions, a small degree of instruction-level parallelism, and multiple cores on a chip. In addition, higher performance has been obtained by using a network of such PC or server processors both for

---

[13]W. J. Bouknight, Stewart A. Denenberg, David F. McIntyre, J.M. Randal, Amed H. Sameh, and Daniel L. Slotnick, 1972, The Illiac IV system, Proceedings of the IEEE 60(4): 369-388.

scientific computing and to serve an aggregate workload of independent tasks, such as Web services. The recent graphical-processing-unit chips also borrow ideas from the body of work on parallel hardware.

As noted previously, it has long been clear that one of the major hurdles in parallel computing is software development. Even if there were sufficient and appropriate software abstractions to enable parallel programming (Google's MapReduce, discussed below, is an example of a successful approach for a particular class of problems), characteristics of the application under consideration can still pose challenges. To exploit parallelism successfully, several things are necessary:

- The application under consideration must inherently have parallelism. Not all programs are amenable to parallelization, but many computationally intensive problems have high-level tasks that are largely independent or they are processing large datasets in which the operations on each individual item are mostly independent. Scientific simulations and graphics applications, for example, often have substantial parallelism to exploit because they perform operations on large arrays of data. Web servers process requests for a large set of users that involve mostly independent operations.
- Assuming that the application under consideration has sufficient parallelism, the parallelism must be identified. Either the programmer explicitly specifies the parallel tasks when developing the application or the system needs to infer the parallelism and automatically take advantage of it. If the parallelism involves tasks that are not entirely independent, the programmer or system also needs to identify communication and synchronization between tasks.
- Efficiency needs to be taken into account, inasmuch as it is not unusual for an initial parallel implementation to run more slowly than its serial counterpart. Parallelism inevitably incurs overhead costs, which include the time to create parallelism and to communicate and synchronize between parallel components. Some applications do not divide neatly into tasks that entail equal amounts of work, so the load must be balanced and any overhead associated with load-balancing managed. Locality is no longer a question of working within a single memory hierarchy, but one of managing the distribution of data between parallel tasks. It is important that multiprocessors exploit coarse-grain parallelism to minimize synchronization and communication overhead and exploit locality. It is this phase that naturally turns programmers

- into performance engineers, making them more aware of all performance issues in the application.
- Last, but definitely not least, the parallel program must be correct. Parallelism introduces a new class of errors due to the creation of parallel computations for work that is not independent or to failure to communicate or synchronize correctly between parallel tasks. Parallelism also introduces new problems into testing and debugging, in that program behavior can depend on the schedule of execution of different processes. Those dependences make it difficult to test programs thoroughly and to reproduce faulty behavior when it is observed. Parallel programming approaches can be restricted to eliminate some of the problems by requiring, for example, that programs communicate only through synchronous messaging or that a compiler verify the independence of loop iterations before running them in parallel. But those approaches can limit the effectiveness of parallel computing by adding overhead or restricting its use. Writing correct sequential code is hard enough, but the complexity of parallel programming is so high that only a small percentage of the programmers in the industry today are competent at it.

The software industry has invested a lot of time and effort in creating the existing software base. In the past, when growth in computing performance was on its exponentially rising curve (see previous chapters), most applications would automatically run faster on a faster machine.

There has been a lot of research to try to minimize the cost of software development for parallel machines. There will be a major prize if we succeed in doing it in a way that allows reuse of the large software-code base that has been developed over many years. Automatic parallelization has some successes, such as instruction-level parallelism and fine-grained loop-level parallelism in FORTRAN programs operating on arrays. The theory for automatically transforming code of this sort is well understood, and compilers often rely on substantial code restructuring to run effectively. In practice, the performance of the programs is quite sensitive to the particular details of how the program is written, and these approaches are more effective in fine-grained parallelism than in the more useful coarse-grained parallelism. However, there has not been sufficient demand in the parallel-software tool industry to sustain research and development.

Most programs, once coded sequentially, have many data dependences that prevent automatic parallelization. Various studies that analyzed the inherent dependences in a sequential program have found a lot of data dependences in such programs. Sometimes, a data dependence is a result of a reuse of memory locations, which may be eliminated through analysis. It is perhaps not surprising that programs written with a sequential-machine model cannot automatically be parallelized. Researchers have also explored whether expressing computation in a different way may expose the parallelism inherent in a program more readily. In data-flow and functional programs, the memory is not reused, and computation can proceed as soon as the operands are ready. That translates to abundant parallelism but adds substantial cost in memory use and copying overhead because data structures cannot be updated in place. Analysis is then necessary to determine when memory can be reused, which is the case if the program will no longer touch the old structure. Optimizing a functional language then becomes a problem of replacing the creation of new data structures with in-place updates of old ones. The analysis requires the discovery of all potential read accesses to a data structure before it can be reclaimed, which in turn necessitates analyzing aliases to detect whether two expressions can refer to the same value. Such analysis is no easier than automatic parallelization in that both require accurate aliasing information, which is not practical for problems with complex pointer-based data structures.

THE STATE OF THE ART OF PARALLEL PROGRAMMING

Notwithstanding the challenges presented by parallelism, there have been some success stories over the years. This section describes several parallel approaches to illustrate the array of applications to which parallelism has been applied and the array of approaches that are encompassed under the term *parallelism*. None of the approaches described here constitutes a general-purpose solution, and none meets all the emerging requirements (described above) that the performance slowdown and new architectures will require; but they may offer lessons for moving forward. Historically, the success of any given programming approach has been strongly influenced by the availability of hardware that is well matched to it. Although there are cases of programming systems that run on hardware that is not a natural fit, the trends in parallel hardware have largely determined which approaches are successful. The specific examples discussed here are thread programming for shared memory, message-passing interface, MapReduce (used to exploit data parallelism and distributed computation), and ad hoc distributed computation (as in such efforts as SETI@home). The latter is not normally thought of as a parallel-programming approach, but it is offered here to demonstrate the variety of approaches that can be considered parallelism.

Thread Programming for Shared Memory

The concept of independent computations within a shared-memory space as threads is popular for programming of parallel shared-memory systems and for writing applications that involve asynchronous interaction with the environment—for example, user interfaces in which one thread of a computation is waiting for a response from the user while another thread is updating a display and a third may be performing calculations on the basis of earlier input. In the latter case, there may be only a single processor in the system and therefore no real parallelism, but the thread execution is interleaved, making the computations appear to be concurrent. And the performance advantage is real, in the same sense that allowing someone with only one item to go ahead in a supermarket line can result in a net "system throughput" for all concerned. The word *thread* is used in a variety of ways in different programming systems, but in general two properties are associated with threads: the ability to create parallel work dynamically, so the number of threads in a given execution may vary over time; and the ability of threads to read and write shared variables.

Threads require little or no language modification but only a small set of primitive features to create and destroy parallelism and synchronization to control access to shared variables. The most common system-level library for threads is the POSIX Thread, or "PThread" library, which allows a programmer to create a parallel computation by providing a function and an argument that will be passed to that function when it begins executing. Threads are first-class values in the language, so they can be named, stored in data structures, and passed as arguments, and one can wait for completion of a thread by performing a "join" operation on the thread. The PThread library contains synchronization primitives to acquire and release locks, which are used to give one thread exclusive access to shared data structures. There are other features of thread creation and synchronization, but the set of primitives is relatively small and easy to learn.

Although the set of primitives in PThreads is small, it is a low-level programming interface that involves function pointers, loss of type information on the arguments, and manual error-checking. To address those issues, there are several language-level versions of threads that provide a more convenient interface for programmers. For example, the Java thread model and more recent Thread Building Blocks (TBB) library for C++ use object-oriented programming abstractions to provide thread-management capabilities in those languages. Java threads are widely use for programming user interfaces and other concurrent programming problems, as described above, but the runtime support for true parallelism is more recent, so there is less experience in using Java threads for parallel programming. TBB is also relatively recent but has demonstrated support in both industry and academe. In the 1980s, when shared-memory hardware was especially popular, the functional language community introduced the notion of a "future"[14] that wrapped around a function invocation and resulted in an implicit synchronization of thread completion; any attempt to access the return value of the function would wait until the thread had completed. The closely related idea of a "promise"[15] is also wrapped around a function invocation but uses a special return type that must be explicitly unwrapped, making the wait for thread completion explicit.

Two issues with dynamic thread creation are the overhead of thread creation and the policy for load-balancing threads among processors. A program written to create a thread everywhere that one could be used will typically overwhelm the scheduler. Several research efforts address such problems, including one extension of C called Cilk,[16] which is now supported by the Cilk Arts company. Cilk uses the syntactic block structure of the language to restrict thread creation and completion to simple nested patterns. It also uses a lazy thread creation model, which allows many threads to execute with low overhead as simple function calls if no processor is available to execute the thread. Instead, the runtime system on an idle processor steals work randomly from other processors. Allowing lazy-thread creation affects the semantics of the threading model; the PThread semantics require that each thread eventually execute even if there are enough other threads to keep the processors busy. In contrast, Cilk makes no such guarantee, so in a Cilk program, if one thread waits for a variable to be set by another thread, it may wait forever. Waiting for a variable to be set or a data structure to be updated without some explicit synchronization is generally considered dubious programming practice in parallel code although it is a popular technique for avoiding the overhead associated with system-provided synchronization primitives.

In scientific computing, the most popular programming interface for shared-memory programming is OpenMP, a standard that emphasizes loop-level parallelism but also has support for more general task parallelism. OpenMP addresses the thread-overhead issue by dividing a set of iterations into groups so that each thread handles a set of iterations, and the programmer is able to control the load-balancing policy. It also gives more flexibility in restricting data that are private to a thread, as opposed to allowing them to be shared by all threads; and controlled forms of syn-

---

[14] Robert Halstead, 1985, MULTILISP: A language for concurrent symbolic computation, ACM Transactions on Programming Languages and Systems 7(4): 501-538.

[15] Barbara Liskov, 1998, Distributed programming in Argus, Communications of the ACM 31(3): 300-312.

[16] For more on Cilk, see its project page at The Cilk Project, MIT website, at http://supertech.csail.mit.edu/cilk/index.html.

chronization and parallelism avoid some kinds of programming errors. OpenMP is sometimes used with message-passing to create a hybrid programming model: large-scale computing clusters in which individual nodes have hardware support for shared memory.

The biggest drawbacks to thread programming are the potential for uncontrolled access to shared variables and the lack of locality control. Shared-variable access results in race conditions, in which two threads access a variable and at least one writes the variable; this results in indeterminate behavior that depends on the access order and may vary from one run to the next. Those accesses make testing and debugging especially difficult. Synchronization primitives—such as locks, which are used to avoid races—have their own forms of subtle errors: threads acquiring multiple locks can form deadlocks in which each of two threads are waiting for a lock held by the other thread. Some tools have been developed by the research community to detect those kinds of parallel-programming errors, but they have not reached the level of generality, accuracy, and speed that would encourage widespread deployment. Thread programming remains an error-prone process best handled by expert programmers, not the broader programming community of persons who have little formal training in programming, who would find it extremely challenging to create and maintain reliable code with these models. The broader community fueled the growth in computing applications and the associated economic and social effects.

The lack of locality support in threaded models limits the scalability of the underlying architecture and calls for some form of cache coherence, which traditionally has been a hardware challenge that grows exponentially harder as the number of processors grows.[17] On the scale of chip multiprocessor systems available today, the problem is tractable, but even with on-chip data transfer rates, it is unclear how performance will be affected as core counts grow. Further complicating the programming problem for shared memory, many shared-memory machines with coherent caches use a relaxed consistency model; that is, some memory operations performed by one thread may appear to be performed in a different order by another thread. There is some research on mapping OpenMP and Cilk to distributed-memory systems or building shared-memory support with cache coherence in software, but the locality-aware model has often proved superior in performance even on systems with hardware-supported shared memory. Savvy thread programmers will use system mechanisms to control data layout and thread affinity to processors, but

---

[17]More recent parallel languages, such as Chapel and X10, have explicitly included support for locality.

in the end, this model is best reserved for a relatively small set of compiler writers, runtime-system developers, and low-level library programmers.

Message-Passing Interface

The combination of scalability limits of shared-memory architectures and the cost and performance benefits of building parallel machines from commodity processors made distributed-memory multiprocessors a popular architecture for high-end parallel computing. Those systems can vary from generic clusters built from personal computers and an Ethernet network to more specialized supercomputers with low-latency high-bandwidth networks that are more closely integrated into the processor nodes. As this architectural model become dominant in the early 1990s, several message-passing systems were developed by scientific programming communities, computer scientists, and industry. In the early 1990s, a group with representatives from those communities began a process to specify the message-passing interface (MPI). MPI has since emerged as the de facto standard programming model for high-performance computing and has nearly ubiquitous support among machines, including open-source implementations, such as MPICH and OpenMPI, that can be ported to new interconnection networks with modest effort.[18] Although the standardization decisions in defining MPI were far from obvious, the relative ease with which application developers moved code from one of the previous models to MPI reflects the commonality of base concepts that already existed in the community. MPI has also proved to be a highly scalable programming model and is used today in applications that regularly run on tens of thousands of processor cores, and sometimes over 100,000, in the largest supercomputers in the world. MPI is generally used to program a single cluster or supercomputer that resides at one site, but "grid-computing" variations of MPI and related libraries support programming among machines at multiple sites. It is also low-level programming and has analogues to the challenges presented by machine-language programming mentioned earlier.

MPI has enabled tremendous scientific breakthroughs in nearly every scientific domain with some of the largest computations in climate change, chemistry, astronomy, and various aspects of defense. Computer simulations have demonstrated human effects on climate change and are critical

---

[18]For more on the MPI standard, see the final version of the draft released in May 1994, available online at http://www.mcs.anl.gov/Projects/mpi/standard.html. See also Open MPI: Open source high performance computing at http://www.open-mpi.org. and Peter Pacheco, 1997, Parallel Programming with MPI, fourth edition, San Francisco, Cal.: Morgan Kaufmann.

for international environmental-policy negotiations, as recognized in the 2007 Nobel prize awarded to the Intergovernmental Panel on Climate Change. The climate-modeling problem is far from solved as researchers attempt to identify particular phenomena, such as the disappearance of polar ice caps; effects on the frequency or severity of hurricanes, floods, or droughts; and the effects of various mitigation proposals. The size and accuracy of the computations continue to push the limits of available computing systems, consuming tens of millions of processor-hours each year. The Community Climate Simulation Model and nearly all other major codes used for global-scale long-term climate simulations are written in MPI. A related problem that also relies heavily on MPI is weather-forecasting, which is more detailed but of shorter term and on smaller regional scales than climate models. MPI programs have been used to understand the origins of the universe on the basis of a study of cosmic microwave background (CMB) and to study quantum chromodynamics in particle physics—both uses are based on Nobel prize-wining work in physics. MPI is used in the design of new materials, in crash simulations in the automobile industry, in simulating earthquakes to improve building designs and standards, and in fluid-dynamics simulations for improving aircraft design, engine design, and biomedical simulations.

There are limitations, however. MPI is designed for comparatively coarse-grained parallelism—among clusters of computers—not for parallelism between cores on a chip. For example, most supercomputers installed in the last few years have dual-core or quad-core processing chips, and most applications use an MPI process per core. The model is shifting as application scientists strive to make more effective use of shared-memory chip multiprocessors by exploiting a mixture of MPI with OpenMP or PThreads. MPI-3 is a recent effort to implement MPI on multicore processors although memory-per-core constraints are still a barrier to MPI-style weak scaling. Indeed, the motivation to mix programming models is often driven more by memory footprint concerns than by performance itself because the shared-memory model exploits fine-grained parallelism better in that it requires less memory per thread. Moreover, there is a broad class of applications of particular interest to the defense and intelligence communities for which MPI is not appropriate, owing to its mismatch with the computational patterns of particular problems.

MapReduce: Exploiting Data Parallelism and Distributed Computation

MapReduce is a data-processing infrastructure[19] developed internally by Google and later popularized in the Hadoop open-source version.[20] MapReduce is targeted at large-scale data-parallel workloads in which the input is represented as a set of key-value pairs and computation is expressed as a sequence of two user-provided functions: map and reduce. The Map function processes the input to create an intermediate key-value pair. Intermediate pairs with the same key are aggregated by the system and fed to the reduce function, which produces the final output.

What makes MapReduce particularly compelling is that it frees the programmer from the need to worry about much of the complexity required to run large-scale computations. The programmer needs only to produce the body of the two MapReduce methods, and the system takes care of parallelization, data distribution, fault tolerance, and load-balancing.

The class of problems that can be mapped to this relatively simple processing framework is surprisingly broad. MapReduce was conceived to simplify the data-processing involved in creating a Web index from a set of crawled documents, but developers have also used it for large-scale graph algorithms (for example, finding citation occurrences among scientific articles), computing query-term popularity (for example, Google Trends[21]), and creating language models for statistical machine translation (for example, finding and keeping a count for every unique sequence of five or more terms in a large corpus), and other applications. Within Google itself, MapReduce's popularity has exploded since its introduction.

MapReduce has also been found to be useful for systems much smaller than the large distributed clusters for which it was designed. Ranger et al. examined the adequacy of the MapReduce framework for multicore and multiprocessor systems[22] and found that it was equally compelling as a programming system for this class of machines. In a set of eight applications both coded with a version of MapReduce (the Phoenix runtime) and hand-coded directly, the authors found that MapReduce-coded ver-

---

[19]See Jeffrey Dean and Sanjay Ghemawat, 2008, MapReduce: Simplified data processing on large clusters, Communications of the ACM 51(1): 107-113, and Micheal Noth, 2006, Building a computing system for the world's information, Invited talk, University of Iowa, IT Tech Forum Presentations, April 20, 2006.
[20]See The Apache Hadoop project, available at http://hadoop.apache.org.
[21]See Google trends, available at http://trends.google.com.
[22]Colby Ranger, Ramanan Raghuraman, Arun Penmetsa, Gary Bradski, Christos Kozyrakis, 2007, Evaluating MapReduce for multi-core and multiprocessor systems, Proceedings of the IEEE 13th International Symposium on High-Performance Computer Architecture, Phoenix, Ariz., February 10-14, 2007.

sions roughly matched the performance of the hand-coded versions for five. The remaining three applications did not fit well with MapReduce's key-value data-stream model, and the hand-coded versions performed significantly better.

Despite MapReduce's success as a programming system beyond its initial application domain and machine-architecture context, it is far from a complete solution for extracting and managing parallelism in general. It remains limited to, for example, batch-processing systems and is therefore not suitable for many on-line serving systems. MapReduce also does not extract implicit parallelism from an otherwise sequential algorithm but instead facilitates the partitioning, distribution, and runtime management of an application that is already essentially data-parallel. Such systems as MapReduce and NVIDIA's CUDA,[23] however, point to a solution strategy for the general programming challenge of large-scale parallel systems. The solutions are not aimed at a single programming paradigm for all possible cases but are based on a small set of programming systems that can be specialized for particular types of applications.

Distributed Computation—Harnessing the World's Spare Computing Capacity

The increasing quality of Internet connectivity around the globe has led researchers to contemplate the possibility of harnessing the unused computing capability of the world's personal computers and servers to perform extremely compute-intensive parallel tasks. The most notable examples have come in the form of volunteer-based initiatives, such as SETI@home (http://setiathome.berkeley.edu/) and Folding@home (http://folding.stanford.edu/). The model consists of breaking a very large-scale computation into subtasks that can operate on relatively few input and result data, require substantial processing over that input set, and do not require much (or any) communication between subtasks other than passing of inputs and results. Those initiatives attract volunteers (individuals or organizations) that sympathize with their scientific goals to donate computing cycles on their equipment to take on and execute a number of subtasks and return the results to a coordinating server that coalesces and combines final results.

The relatively few success cases of the model have relied not only on the friendly nature of the computation to be performed (vast data parallelism with very low communication requirements for each unit of

---

[23]CUDA is a parallel computing approach aimed at taking advantages of NVIDIA graphical processing units. For more, see CUDA zone, NVIDIA.com, available at http://www.nvidia.com/object/cuda_home.html.

computing) but also on the trust of the volunteers that the code is safe to execute in their machines. For more widespread adoption, this particular programming model would require continuing improvements in secure execution technologies, incorporation of an economic model that provides users an incentive to donate their spare computing capacity, and improvements in Internet connectivity. To some extent, one can consider large illegitimately assembled networks of hijacked computers (or botnets) to be an exploitation of this computing model; this exemplifies the potential value of harnessing a large number of well-connected computing systems toward nobler aims.

Summary Observations

These success stories show that there are already a wide variety of computational models for parallel computation and that science and industry are successfully harnessing parallelism in some domains. The parallelism success stories bode well for the future if we can find ways to map more applications to the models or, for computations that do not map well to the models, if we can develop new models.

PARALLEL-PROGRAMMING SYSTEMS AND THE PARALLEL SOFTWARE "STACK"

The general problem of designing parallel algorithms and programming them to exploit parallelism is an extremely important, timely, and unsolved problem. The vast majority of software in use today is sequential. Although the previous section described examples of parallel approaches that work in particular domains, general solutions are still lacking. Many successful parallel approaches are tied to a specific type of parallel hardware (MPI and distributed clusters; storage-cluster architecture, which heavily influenced MapReduce; openGL and SIMD graphics processors; and so on). The looming crisis that is the subject of this report comes down to the question of how to continue to improve performance scalability as architectures continue to change and as more and more processors are added. There has been some limited success, but there is not yet an analogue of the sequential-programming models that have been so successful in software for decades.

We know some things about what new parallel-programming approaches will need. A high-level performance-portable programming model is the only way to restart the virtuous cycle described in Chapter 2. The new model will need to be portable over successive generations of chips, multiple architectures, and different kinds of parallel hardware, and it will need to scale well. For all of those goals to be achieved, the entire software stack will need to be rethought, and architectural assumptions will need to be included in the stack. Indeed, in the future, the term *software stack* will be a misnomer. A "software-hardware stack" will be the norm. The hardware, the programming model, and the applications will all need to change.

A key part of modern programming systems is the software stack that executes the program on the hardware. The stack must also allow reasoning about the five main challenges to scalable and efficient performance: parallelism, communication, locality, synchronization, and load-balancing. The components of a modern software stack include

- Libraries: Generic and domain-specific libraries provide application programmers with predefined software components that can be included in applications. Because library software may be reused in many applications, it is often highly optimized by hand or with automated tools.
- Compiler: An ahead-of-time or a just-in-time compiler translates the program into assembly code and optimizes it for the underlying hardware. Just-in-time compilers combine profile data from the current execution with static program analysis to perform optimizations.
- Runtime system or virtual machine: These systems manage fine-grain memory resources, application-thread creation and scheduling, runtime profiling, and runtime compilation.
- Operating system: The operating system manages processes and their resources, including coarse-grain memory management.
- Hypervisors: Hypervisors abstract the hardware context to provide performance portability for operating systems among hardware platforms.

Because programming systems are mostly sequential, the software stack mostly optimizes and manages sequential programs. Optimizing and understanding the five challenges at all levels of the stack for parallel approaches will require substantial changes in these systems and their interfaces, and perhaps researchers should reconsider whether the overall structure of the stack is a good one for parallel systems.

Researchers have made some progress in system support for providing and supporting parallel-programming models.[24] Over the years, researchers and industry have developed parallel-programming system tools, which include languages, compilers, runtime environments, libraries, components, and frameworks to assist programmers and software developers in managing parallelism. We list some examples below.

- Runtime abstractions: multiprogramming and virtualization. The operating system can exploit chip multiprocessors at a coarse granularity immediately because operating systems can run multiple user and kernel processes in parallel. Virtualization runs multiple operating systems in parallel. However, it remains challenging to manage competition for shared resources, such as caches, when a particular application load varies dramatically.
- Components: database transactions and Web applications. Database-transaction systems provide an extremely effective abstraction in which programs use a sequential model, without the need to worry about synchronization and communication, and the database coordinates all the parallelism between user programs. Success in this regard emerged from over 20 years of research in parallelizing database systems.
- Frameworks: three-tiered Web applications and MapReduce. Such frameworks as J2EE and Websphere make it easy to create large-scale parallel Web applications. For example, MapReduce (described above) simplifies the development of a large class of distributed applications that combine the results of the computation of distributed nodes. Web applications follow the database-transaction model in which users write sequential tasks and the framework manages the parallelism.
- Libraries: graphics libraries. Graphics libraries for DirectX 10 and OpenGl hide the details of parallelism in graphics hardware from the user.
- Languages: Cuda Fortress, Cilk, x10, and Chapel. These languages seek to provide an array of high-level and low-level abstractions that help programmers to develop classes of efficient parallel software faster.

What those tools suggest is that managing parallelism is another, more

---

[24]One recent example was a parallel debugger, STAT, from the Lawrence Livermore National Laboratory, available at http://www.paradyn.org/STAT/STAT.html, presented at Supercomputing 2008. See Gregory L. Lee, Dong H. Ahn, Dorian C. Arnold, Bronis R. de Supinski, Matthew Legendre, Barton P. Miller, Martin Schulz, and Ben Liblit, 2008, Lessons learned at 208K: Towards debugging millions of cores, available online at ftp://ftp.cs.wisc.edu/paradyn/papers/Lee08ScalingSTAT.pdf, last accessed on November 8, 2010.

challenging facet of software engineering—it can be thought of as akin to a complex version of the problem of resource management. Parallel-program productivity can be improved if we can develop languages that provide useful software-engineering abstractions for parallelism, parallel components and libraries that programmers can reuse, and a software-hardware stack that can facilitate reasoning about all of them.

MEETING THE CHALLENGES OF PARALLELISM

The need for robust, general, and scalable parallel-software approaches presents challenges that affect the entire computing ecosystem. There are numerous possible paths toward a future that exploits abundant parallelism while managing locality. Parallel programming and parallel computers have been around since the 1960s, and much progress has been made. Much of the challenge of parallel programming deals with making parallel programs efficient and portable without requiring heroic efforts on the part of the programmer. No subfield or niche will be able to solve the problem of sustaining growth in computing performance on its own. The uncertainty about the best way forward is inhibiting investment. In other words, there is currently no parallel-programming approach that can help drive hardware development. Historically, a vendor might have taken on risk and invested heavily in developing an ecosystem, but given all the uncertainty, there is not enough of this investment, which entails risk as well as innovation. Research investment along multiple fronts, as described in this report, is essential.[25]

Software lasts a long time. The huge entrenched base of legacy software is part of the reason that people resist change and resist investment in new models, which may or may not take advantage of the capital investment represented by legacy software. Rewriting software is expensive. The economics of software results in pressure against any kind of innovative models and approaches. It also explains why the approaches we have seen have had relatively narrow applications or been incremental. Industry, for example, has turned to chip multiprocessors (CMPs) that replicate existing cores a few times (many times in the future). Care is taken to maintain backward compatibility to bring forward the existing multi-billion-dollar installed software base. With prospects dim for

---

[25] A recent overview in Communications of the ACM articulates the view that developing software for parallel cores needs to become as straightforward as writing software for traditional processors: Krste Asanovic, Rastislav Bodik, James Demmel, Tony Keaveny, Kurt Keutzer, John Kubiatowicz, Nelson Morgan, David Patterson, Koushik Sen, John Wawrzynek, David Wessel, and Katherine Yelick, 2009, A view of the parallel computing landscape, Communications of the ACM 52(10): 56-67, available online at http://cacm.acm.org/magazines/2009/10/42368-a-view-of-the-parallel-computing-landscape/fulltext.

repeated doublings of single-core performance, CMPs inherit the mantle as the most obvious alternative, and industry is motivated to devote substantial resources to moving compatible CMPs forward. The downside is that the core designs being replicated are optimized for serial code with support for dynamic parallelism discovery, such as speculation and out-of-order execution, which may waste area and energy for programs that are already parallel. At the same time, they may be missing some of the features needed for highly efficient parallel programming, such as lightweight synchronization, global communication, and locality control in software. A great deal of research remains to be done on on-chip networking, cache coherence, and distributed cache and memory management.

One important role for academe is to explore CMP designs that are more aggressive than industry's designs. Academics should project both hardware and software trends much further into the future to seek possible inflection points even if they are not sure when or even whether transitioning a technology from academe to industry will occur. Moreover, researchers have the opportunity to break the shackles of strict backward compatibility. Promising ideas should be nurtured to see whether they can either create enough benefit to be adopted without portability or to enable portability strategies to be developed later. There needs to be an intellectual ecosystem that enables ideas to be proposed, cross-fertilized, and refined and, ultimately, the best approaches to be adopted. Such an ecosystem requires sufficient resources to enable contributions from many competing and cooperating research teams.

Meeting the challenges will involve essentially all aspects of computing. Focusing on a single component—assuming a CMP architecture or a particular number of transistors, focusing on data parallelism or on heterogeneity, and so on—will be insufficient to the task. Chapter 5 discusses recommendations for research aimed at meeting the challenges.

5

Research, Practice, and Education to Meet Tomorrow's Performance Needs

Early in the 21st century, single-processor performance stopped growing exponentially, and it now improves at a modest pace, if at all. The abrupt shift is due to fundamental limits on the power efficiency of complementary metal oxide semiconductor (CMOS) integrated circuits (used in virtually all computer chips today) and apparent limits on what sorts of efficiencies can be exploited in single-core architectures. A sequential-programming model will no longer be sufficient to facilitate future information technology (IT) advances.

Although efforts to advance low-power technology are important, the only foreseeable way to continue advancing performance is with parallelism. To that end, the hardware industry recently began doubling the number of cores per chip rather than focusing solely on more performance per core and began deploying more aggressive parallel options, for example, in graphics processing units (GPUs). Attaining dramatic IT advances in the future will require programs and supporting software systems that can access vast parallelism. The shift to explicitly parallel hardware will fail unless there is a concomitant shift to useful programming models for parallel hardware. There has been progress in that direction: extremely skilled and savvy programmers can exploit vast parallelism (for example, in what has traditionally been referred to as high-performance computing), domain-specific languages flourish (for example, SQL and DirectX), and powerful abstractions hide complexity (for example, MapReduce). However, none of those developments comes close to the ubiquitous support for programming parallel hardware that is required to sustain growth in computing performance and meet society's expectations for IT. (See Box 5.1 for additional context regarding other aspects of computing research that should not be neglected while the push for parallelism in software takes place.)

The findings and results described in this report represent a serious set of challenges not only for the computing industry but also for the many sectors of society that depend on advances in IT and computation. The findings also pose challenges to U.S. competitiveness: a slowdown in the growth of computing performance will have global economic and political repercussions. The committee has developed a set of recommended actions aimed at addressing the challenges, but the fundamental power and energy constraints mean that even our best efforts may not offer a complete solution. This chapter presents the committee's recommendations in two categories: research—the best science and engineering minds must be brought to bear; and practice—how we go about developing computer hardware and software *today* will form a foundation for future performance gains. Changes in education are also needed; the emerging generation of technical experts will need to understand quite different (and in some cases not yet developed) models of thinking about IT, computation, and software.

SYSTEMS RESEARCH AND PRACTICE

Algorithms

In light of the inevitable trend toward parallel architectures and emerging applications, one must ask whether *existing applications are amenable algorithmically for decomposition on any parallel architecture*. Algorithms based on context-dependent state machines are not easily amenable to parallel decomposition. Applications based on those algorithms have always been around and are likely to gain more importance as security needs grow. Even so, there is a large amount of throughput parallelism in these applications, in that many such tasks usually need to be processed simultaneously by a data center.

At the other extreme, there are applications that have obvious parallelism to exploit. The abundance of parallelism in a vast majority of those underlying algorithms is *data-level parallelism*. One simple example of data-level parallelism for mass applications is found in two-dimensional (2D) and three-dimensional (3D) media-processing (image, signal, graphics, and so on), which has an abundance of primitives (such as blocks, triangles, and grids) that need to be processed simultaneously. Continuous growth in the size of input datasets (from the text-heavy Internet of the past to 2D-media-rich current Internet applications to emerging 3D

> **BOX 5.1**
> **React, But Don't Overreact, to Parallelism**
>
> As this report makes clear, software and hardware researchers and practitioners should address important concerns regarding parallelism. At such critical junctures, enthusiasm seems to dictate that all talents and resources be applied to the crisis at hand. Taking the longer view, however, a prudent research portfolio must include concomitant efforts to advance all systems aspects, lest they become tomorrow's bottlenecks or crises.
>
> For example, in the rush to innovate on chip multiprocessors (CMPs), it is tempting to ignore sequential core performance and to deploy many simple cores. That approach may prevail, but history and Amdahl's law suggest caution. Three decades ago, a hot technology was vectors. Pioneering vector machines, such as the Control Data STAR-100 and Texas Instruments ASC, advanced vector technology without great concern for improving other aspects of computation. Seymour Cray, in contrast, designed the Cray-1[1] to have great vector performance as well as to be the world's fastest scalar computer. Ultimately, his approach prevailed, and the early machines faded away.
>
> Moreover, Amdahl's law raises concern.[2] Amdahl's limit argument assumed that a fraction, P, of software execution is infinitely parallelizable without overhead, whereas the remaining fraction, 1 - P, is totally sequential. From that assumption, it follows that the speedup with N cores—execution time on N cores divided by execution time on one core—is governed by $1/[(1 - P) + P/N]$. Many learn that equation, but it is still instructive to consider its harsh numerical consequences. For N = 256 cores and a fraction parallel P = 99%, for example, speedup is bounded by 72. Moreover, Gustafson[3] made good "weak scaling" arguments for why some software will fare much better. Nevertheless, the committee is skeptical that most future software will avoid sequential bottlenecks. Even such a very parallel approach as MapReduce[4] has near-sequential activity as the reduce phase draws to a close.
>
> For those reasons, it is prudent to continue work on faster sequential cores,

Internet applications) has been important in the steady increase in available parallelism for these sorts of applications.

A large and growing collection of applications lies between those extremes. In these applications, there is parallelism to be exploited, but it is not easy to extract: it is less regular and less structured in its spatial and temporal control and its data-access and communication patterns. One might argue that these have been the focus of the high-performance computing (HPC) research community for many decades and thus are well understood with respect to those aspects that are amenable to parallel decomposition. The research community also knows that algorithms best suited for a serial machine (for example, quicksort, simplex, and gaston) differ from their counterparts that are best suited for parallel machines especially with an emphasis on energy efficiency (for example, on large-content addressable-memory structures) and perhaps on-demand scaling (to be responsive to software bottlenecks). Hill and Marty[5] illuminate some potential opportunities by extending Amdahl's law with a corollary that models CMP hardware. They find, for example, that as Moore's law provides more transistors, many CMP designs benefit from increasing the sequential core performance and considering asymmetric (heterogeneous) designs where some cores provide more performance (statically or dynamically).

Finally, although the focus in this box is on core performance, many other aspects of computer design continue to require innovation to keep systems balanced. Memories should be larger, faster, and less expensive. Nonvolatile storage should be larger, faster, and less expensive and may merge with volatile memory. Networks should be faster (higher bandwidth) and less expensive, and interfaces to networks may need to get more closely coupled to host nodes. All components must be designed for energy-efficient operation and even more energy efficiency when not in current use.

---

[1] Richard M. Russell, 1978, The Cray-1 computer system, Communications of the ACM 21(1): 63-72

[2] Gene M. Amdahl, 1967, Validity of the single-processor approach to achieving large scale computing capabilities, AFIPS Conference Proceedings, Atlantic City, N.J,, April 18-20, 1967, pp. 483-485.

[3] John L. Gustafson, 1998, Reevaluating Amdahl's law, Communications of the ACM 31(5): 532-533.

[4] Jeffrey Dean and Sanjay Ghemawat, 2004, MapReduce: Simplified data processing on large clusters, Symposium on Operating System Design and Implementation, San Francisco, Cal., December 6-8, 2004.

[5] Mark D. Hill and Michael R. Marty, 2008, Amdahl's law in the multicore era, IEEE Computer 41(7): 33-38, available online at http://www.cs.wisc.edu/multifacet/papers/tr1593_amdahl_multicore.pdf.

(for example, mergesort, interior-point, and gspan). Given the abundance of single-thread machines in mass computing, commonly found implementations of these algorithms on mass machines are almost always the nonparallel or serial-friendly versions. Attempts to extract parallelism from the serial implementations are unproductive exercises and likely to be misleading if they cause one to conclude that the original problem has an inherently sequential nature. Therefore, there is an opportunity to benefit from the learning and experience of the HPC research and to reformulate problems in terms amenable to parallel decomposition.

Three additional observations are warranted in the modern context of data-intensive connected computing:

- In the growing segment of the entertainment industry, in contrast with the scientific computing requirements of the past, approximate or sometimes even incorrect solutions are often good enough if end users are insensitive to the details. An example is cloud simulation for gaming compared with cloud simulation for weather prediction. Looser correctness requirements almost always make problems more amenable to parallel algorithms because strict dependence requires synchronized communication, whereas an approximation often can eliminate communication and synchronization.
- The serial fraction of any parallel algorithm would normally dominate the performance—a manifestation of Amdahl's law (described in Box 2.2) typically referred to as weak scaling. That is true for a fixed problem size. However, if the problem size continues to scale, one would observe continuously improved performance scaling of a parallel architecture, provided that it could effectively handle the larger data input. This is the so-called Gustafson corollary[1] to Amdahl's law. Current digitization trends are leading to input-dataset scaling for most applications (for example, today there might be 1,000 songs on a typical iPod, but in another couple of years there may be 10,000).
- Massive, easily accessible real-time datasets have turned some previous sparse input into much denser inputs. This has at least two important algorithmic implications: the problem becomes more regular and hence more amenable to parallelism, and better training and hence better classification accuracies make additional parallel formulations usable in practice. Examples include scene completion in photographs[2] and language-neutral translation systems.[3]

For many of today's applications, the underlying algorithms in use do not assume or exploit parallel processing explicitly, except as in the cases described above. Instead, software creators typically depend, implicitly or explicitly, on compilers and other layers of the programming environment to parallelize where possible, leaving the software developer free to think sequentially and focus on higher-level issues. That state of affairs will need to change, and a fundamental focus on parallelism will be needed in

---

[1] John L. Gustafson, 1998, Reevaluating Amdahl's law, Communications of the ACM 31(5): 532-533.
[2] See James Hays and Alexei A. Efros, 2008, Scene completion using millions of photographs, Communications of the ACM 51(10): 87-94.
[3] See Jim Giles, 2006, Google tops translation ranking, Nature.com, November 7, 2006, available online at http://www.nature.com/news/2006/061106/full/news061106-6.html.

designing solutions to specific problems in addition to general programming paradigms and models.

Recommendation: Invest in research in and development of algorithms that can exploit parallel processing.

Programming Methods and Systems

Many of today's programming models, languages, compilers, hypervisors, and operating systems are targeted primarily at single-core hardware. For the future, all these layers in the stack must explicitly target parallel hardware. The intellectual keystone of this endeavor is rethinking programming models. Programmers must have appropriate models of computation that express application parallelism in such a way that diverse and evolving computer hardware systems and software can balance computation and minimize communication among multiple computational units. There was a time in the late 1970s when even the conventional sequential-programming model was thought to be an eventual limiter of software creation, but better methods and training largely ameliorated that concern. We need advances in programmer productivity for parallel systems similar to the advances brought first by structured programming languages, such as Fortran and C, and then later by managed programming languages, such as C# and Java.

The models themselves may or may not be explicitly parallel; it is an open question whether and when most programmers should be exposed to explicit hardware parallelism. The committee does not call for a singular programming model, because a unified solution may or may not exist. Instead, it recommends the exploration of alternative models—perhaps domain-specific—that can serve as sources of possible future unification. Moreover, the committee expects that some programming models will favor ease of use by a broad base of programmers who are not necessarily expert whereas others will target expert programmers who seek the highest performance for critical subsystems that get extensively reused.

Additional research is needed in the development of new libraries and new programming languages that embody the new programming models described above. Development of such libraries will facilitate rapid prototyping of complementary and competing ideas. The committee expects that some of the languages will be easier for more typical programmers to use—that is, they will appear on the surface to be sequential or declarative—and that others will target efficiency and, consequently, expert programmers.

New programming languages—especially those whose abstractions are far from the underlying parallel hardware—will require new compilation and runtime support. Fortress, Chapel, and X10 are three new recently proposed general-purpose parallel languages, but none of them has yet developed a strong following.[4] Experience has shown that it is generally exceedingly difficult to parallelize sequential code effectively—or even to parallelize and redesign highly sequential algorithms. Nevertheless, we must redouble our efforts on this front in part by changing the languages, targeting specific domains, and enlisting new hardware support.

We also need more research in system software for highly parallel systems. Although the hypervisors and operating systems of today can handle some modest parallelism, future systems will include many more cores (and multithreaded contexts), whose allocation, load-balancing, and data communication and synchronization interactions will be difficult to handle well. Solving those problems will require a rethinking of how computation resources are viewed, much as increased physical memory size led to virtual memory a half-century ago.

Recommendation: Invest in research in and development of programming methods that will enable efficient use of parallel systems not only by parallel systems experts but also by typical programmers.

Computer Architecture and Hardware

Most 20th-century computers used a single sequential processor, but many larger computers—hidden in the backroom or by the Internet—harnessed multiple cores on separate chips to form a symmetric multiprocessor (SMP). When industry was unable to use more transistors on a chip for a faster core effectively, it turned, by default, to implementing multiple cores per chip to provide an SMP-like software model. In addition, special-purpose processors—notably GPUs and digital signal processing (DSP) hardware—exploited parallelism and were very successful in important niches.

Researchers must now determine the best way to spend the transistor bounty still provided by Moore's law.[5] On the one hand, we must examine and refine CMPs and associated architectural approaches. But CMP architectures bring numerous issues to the fore. Will multiple cores work in most computer deployments, such as in desktops and even in mobile phones? Under what circumstances should some cores be more capable

---

[4]For more on Fortress, see the website of Project Fortress community, at http://project fortress.sun.com/Projects/Community. For more on Chapel, see the website The Chapel parallel programming language, at http://chapel.cray.com. For more on X10, see the website The X10 programming language, at http://x10.codehaus.org/.

[5]James Larus, 2009, Spending Moore's dividend, Communications of the ACM 52(5): 62-69.

than others or even use different instruction-set architectures? How can cores be harnessed together temporarily in an automated or semiautomated fashion to overcome sequential bottlenecks? What mechanisms and policies will best exploit locality and ease communication? How should synchronization and scheduling be handled? How will challenges associated with power and energy be addressed? What do the new architectures mean for system-level features, such as reliability and security?

Research in computer architecture must focus on providing useful, programmable systems driven by important applications. It is well known that customizing hardware for a specific task yields more efficient and higher-performance hardware. DSP chips are one example. Research is needed to understand application characteristics to drive parallel-hardware design. There is a bit of a chicken-and-egg problem. Without effective CMP hardware, it is hard to motivate programmers to build parallel applications; but it is also difficult to build effective hardware without parallel applications. Because of the lack of parallel applications, hardware designers are at risk of developing idiosyncratic CMP hardware artifacts that serve as poor targets for applications, libraries, compilers, and runtime systems. In some cases, progress may be facilitated by domain-specific systems that may lead to general-purpose systems later.

CMPs have now inherited the computing landscape from performance-stalled single cores. To promote robust, long-term growth, however, we need to look for alternatives to CMPs. Some of the alternatives may prove better; some may pioneer improvements in CMPs; and even if no alternative proves better, we would then know that CMPs have withstood the assault of alternatives. The research could eschew conventional cores. It could, for example, view the chip as *a tabula rasa* of billions of transistors, which translates to hundreds of functional units; but the best organization of these units into a programmable architecture is an open question. Nevertheless, researchers must be keenly aware of the need to enable useful, programmable systems. Examples include evolving GPUs for more general-purpose programming, game processors, or computational accelerators used as coprocessors; and exploiting special-purpose, energy-efficient engines at some level of granularity for computations, such as fast Fourier transforms, Codec, or encryption. Other tasks to which increased computational capability could be applied include architectural support for machine learning, communication compression, decompression, encryption, and decryption, and dedicated engines for GPS, networking, human interface, search, and video analytics. Those approaches have potential demonstrated advantages in increased performance and energy efficiency relative to a more conservative CMP approach.

Ultimately, we must question whether the CMP-architecture direction, as currently defined, is a good approach for designing most computers. The current CMP architecture preserves object-code compatibility, the heart of the architectural franchise that keeps such companies as Intel and AMD investing heavily. Despite their motivation and ability to expend resources, if systems with CMP architectures cannot be effectively programmed, an alternative will be needed. Is using homogeneous processors in CMP architectures the best approach, or will computer architectures that include multiple but heterogeneous cores be more effective—for example, a single high-performance but power-inefficient processor for programs that are stubbornly sequential and many power-efficient but lower-performance cores for other applications? Perhaps truly effective parallel hardware needs to follow a model that does not assume shared memory parallelism, instead exploiting single-instruction multiple-data approaches, streaming, dataflow, or other paradigms yet to be invented. Are there other important niches like those exploited by GPUs and DSPs? Alternatively, will cores support more graphics and GPUs support more general-purpose programs, so that the line between the two blurs? And most important, are any of those alternatives sufficient to keep the industry driving forward at a pace that can avoid the challenges described elsewhere?

We may also need to consider fundamentally rethinking the nature of hardware in light of today's near-universal connectivity to the Internet. The trend is likely to accelerate. When Google needed to refine the general Internet search problem, it used the MapReduce paradigm so that it could easily and naturally harness the computational horsepower of a very large number of computer systems. Perhaps an equivalent basic shift in how we think about engineering computer systems themselves ought to be considered.

The slowing of growth in single-core performance provides the best opportunity to rethink computer hardware since the von Neumann model was developed in the 1940s. While a focus on the new research challenges is critical, continuing investments are needed in new computation substrates whose underlying power efficiency promises to be fundamentally better than silicon-based CMOSs. In the best case, investment will yield devices and manufacturing methods—as yet unforeseen—that will dramatically surpass the transistor-based integrated circuit. In the worst case, no new technology will emerge to help solve the problems. It is therefore essential to invest in parallel approaches, as outlined previously, and to do so now. Performance is needed immediately, and society cannot wait the decade or two needed to refine a new technology, which may or may not even be on the horizon. Moreover, even if we discover a groundbreaking new technology, the investment in parallelism would not be wasted, inasmuch as it is very likely that advances in parallelism would exploit new technology as well.[6] Substantial research investment should focus on approaches that eschew conventional cores and develop new experimental structures for each chip's billions of transistors.

Recommendation: Invest in research on and development of parallel architectures driven by applications, including enhancements of chip multiprocessor systems and conventional data-parallel architectures, cost-effective designs for application-specific architectures, and support for radically different approaches.

Computer scientists and engineers manage complexity by separating interface from implementation. In conventional computer systems, the separation is recursive and forms the traditional computing stack: applications, programming language, compiler, runtime and virtual machine environments, operating system, hypervisor, and architecture. The committee has expressed above and in Chapter 4 the need for innovation with reference to that stack. However, some long-term work should focus on whether the von Neumann stack is best for our parallel future. The examination will require teams of computer scientists in many subdisciplines. Ideas may focus on changing the details of an interface (for example, new instructions) or even on displacing a portion of the stack (for example, compiling straight down to field-programmable gate arrays). Work should explore first what is possible and later how to move IT from where it is today to where we want it to be.

Recommendation: Focus long-term efforts on rethinking of the canonical computing "stack"—applications, programming language, compiler, runtime, virtual machine, operating system, hypervisor, and architecture—in light of parallelism and resource-management challenges.

Finally, the fundamental question of power efficiency merits considerable research attention. Chapter 3 explains in great detail the power limitations that we are running up against with CMOS technology. But

---

[6]For example, in the 1930s AT&T could see the limitations of relays and vacuum tubes for communication switches and began the search for solid-state devices. Ultimately, AT&T Bell Labs discovered the solid-state semiconductor transistor, which, after several generations of improvements, became the foundation of today's IT. Even earlier, the breakthrough innovation of the stored-program computer architecture (EDSAC) replacing the patch-panel electronic calculator (ENIAC) changed the fundamental approach to computing and opened the door for the computing revolution of the last 60 years. See Arthur W. Burks, Herman H. Goldstine, and John von Neumann, 1946, Preliminary Discussion of the Logical Design of an Electronic Computing Instrument, Princeton, N.J.: Institute for Advanced Study, available online at http://www.cs.unc.edu/~adyilie/comp265/vonNeumann.html.

the power challenges go beyond chip and architectural considerations and warrant attention at all levels of the computing system. New parallel-programming models and approaches will also have an effect on power needs. Thus, research and development efforts are needed in multiple dimensions, with high priority going to software, then to application-specific devices, and then, as described earlier in this report, to alternative devices.[7]

Recommendation: Invest in research and development to make computer systems more power-efficient at all levels of the system, including software, application-specific approaches, and alternative devices. Such efforts should address ways in which software and system architectures can improve power efficiency, such as by exploiting locality and the use of domain-specific execution units.

The need for power efficiency at the processor level was explored in detail in Chapter 3. That chapter explored the decreasing rate of energy-use reduction by silicon technology as feature sizes decrease. One of the consequences of that trend is a flattening of the energy efficiency of computing devices; that is, a given level of performance improvement from a new generation of devices comes with a higher energy need than was the case in previous generations. The increased energy need has broad implications for the sustainability of computing growth from an economic and environmental perspective. That is particularly true for the kinds of server-class systems that are relied on by businesses and users of cloud-computing services.[8]

If improvements in energy efficiency of computing devices flatten out while hardware-cost improvements continue at near historical rates, there will be a shift in the economic costs of computing. The cost basis for deploying computer servers will change as energy-related costs as a fraction of total IT expenses begin to increase. To some extent, that has already been observed by researchers and IT professionals, and this trend

---

[7] Indeed, a new National Science Foundation science and technology center, the Center for Energy Efficient Electronics Science (ES3), has recently been announced. The press release for the center quotes center Director Eli Yablonovitch: "There has been great progress in making transistor circuits more efficient, but further scientific breakthroughs will be needed to achieve the six-orders-of-magnitude further improvement that remain before we approach the theoretical limits of energy consumption." See Sarah Yang, 2010, NSF awards $24.5 million for center to stem increase of electronics power draw, UC Berkeley News, February 23, 2010, available online at http://berkeley.edu/news/media/releases/2010/02/23_nsf_award.shtml.

[8] For more on data centers, their design, energy efficiency, and so on, see Luiz Barroso and Urs Holzle, 2009, The Datacenter as a Computer: An Introduction to the Design of Warehouse-Scale Machines, San Rafael, Cal.: Morgan & Claypool, available online at http://www.morganclaypool.com/doi/abs/10.2200/S00193ED1V01Y200905CAC006.

is partially responsible for the increased attention being given to so-called green-computing efforts.[9]

The following simple model illustrates the relative weight of two of the main components of IT expenses for large data centers: server-hardware depreciation and electricity consumption. Assume a data center filled mostly with a popular midrange server system that is marketed as a high-efficiency system: a Dell PowerEdge Smart 2950 III. As of December 2008, a reasonable configuration of the system was priced at about US$6,000 and may consume from 208 W (at idle) to 313 W (under scientific workload) with an average consumption estimated at 275 W.[10] When the system is purchased as part of a large order, vendors typically offer discounts of at least 15 percent, bringing the actual cost closer to US$5,000. With servers having an operational lifetime of about 4 years, the total energy used by this server in operation is 9,636 kWh, which translates to US$674.52 if it is using the U.S. average industrial cost of electricity for 2008, US$0.0699/kWh.[11] The typical energy efficiency of data-center facilities can multiply IT power consumption by 1.8-2.0,[12] which would result in an actual electricity cost of running the server of up to about US$1,300.

According to that rough model, electricity costs for the server could correspond to about one-fourth of its hardware costs. If hardware-cost efficiency (performance/hardware costs) continues to improve at historical rates but energy efficiency (performance/electricity costs) stops improving, the electricity costs would surpass hardware costs within 3 years. At that point, electricity use could become a primary limiting factor in the growth of aggregate computing performance. Another implication of such a scenario is that at that point most of the IT expenses would be funding development and innovation not in the computing field but in the energy generation and distribution sectors of the economy, and this

---

[9]See, for example, Maury Wright's article, which examines improving power-conversion efficiency (arguably low-hanging fruit among the suite of challenges that need to be addressed): Maury Wright, 2009, Efficient architectures move sources closer to loads, EE Times Design, January 26, 2009, available online at http://www.eetimes.com/showArticle.jhtml?articleID=212901943&cid=NL_eet. See also Randy H. Katz, 2009, Tech titans building boom, IEEE Spectrum, February 2009, available online at http://www.spectrum.ieee.org/green-tech/buildings/tech-titans-building-boom.

[10]See an online Dell power calculator in Planning for energy requirements with Dell servers, storage, and networking, available online at http://www.dell.com/content/topics/topic.aspx/global/products/pedge/topics/en/config_calculator?c=us&cs=555&l=en&s=biz.

[11]See U.S. electric utility sales at a site of DOE's Energy Information Administration: 2010, U.S. electric utility sales, revenue and average retail price of electricity, available online at http://www.eia.doe.gov/cneaf/electricity/page/at_a_glance/sales_tabs.html.

[12]See the TPC-C executive summary for the Dell PowerEdge 2900 at the Transactions Processing Performance Council Web site, June 2008, PowerEdge 2900 Server with Oracle Database 11g Standard Edition One, available online at http://www.tpc.org/results/individual_results/Dell/Dell_2900_061608_es.pdf.

would adversely affect the virtuous cycle described in Chapter 2 that has propelled so many advances in computing technology.

Energy use could curb the growth in computing performance in another important way: by consuming too much of the planet's energy resources. We are keenly aware today of our planet's limited energy budget, especially for electricity generation, and of the environmental harm that can result from ignoring such limitations. Computing confers an immense benefit on society, but that benefit is offset in part by the resources that it consumes. As computing becomes more pervasive and the full value to society of the field's great advances over the last few decades begins to be recognized, its energy footprint becomes more noticeable.

An Environmental Protection Agency report to Congress in 2007[13] states that servers consumed about 1.5 percent of the total electricity generated in the United States in 2006 and that server energy use had doubled from 2000 to 2006. The same report estimated that under current efficiency trends, server power consumption could double once more from 2006 to 2011—a growth that would correspond to the energy output of 10 new power plants (about 5 GW).[14] An interesting way to understand the effect of such growth rates is to compare them with the projections for growth in electricity generation in the United States. The U.S. Department of Energy estimated that about 87 MW of new summer generation capacity would come on line in 2006-2011—an increase of less than 9 percent in that period.[15]

On the basis of those projections, growth in server energy use is outpacing growth in overall electricity use by a wide margin; server use is expected to grow at about 14 percent a year compared with overall electricity generation at about 1.74 percent a year. If those rates are maintained, server electricity use will surpass 5 percent of the total U.S. generating capacity by 2016.

The net environmental effect of more (or better) computing capabilities goes beyond simply accounting for the resources that server-class

---

[13] See the Report to Congress of the U.S. Environmental Protection Agency (EPA) on the Energy Star Program (EPA, 2007, Report to Congress on Server and Data Center Energy Efficiency Public Law 109-431, Washington, D.C.: EPA, available online at http://www.energystar.gov/ia/partners/prod_development/downloads/EPA_Datacenter_Report_Congress_Final1.pdf).

[14] An article in the EE Times suggests that data-center power requirements are increasing by as much as 20 percent per year. See Mark LaPedus, 2009, Green-memory movement takes root, EE Times, May 18, 2009, available online at http://www.eetimes.com/showArticle.jhtml?articleID=217500448&cid=NL_eet.

[15] Find information on DOE planned nameplate capacity additions from new generators at DOE, 2010, Planned nameplate capacity additions from new generators, by energy source, available online at http://www.eia.doe.gov/cneaf/electricity/epa/epat2p4.html.

computers consume. It must also include the energy and emission savings that are enabled by additional computing capacity. A joint report by The Climate Group and the Global e-Sustainability Initiative (GeSI) states that although the worldwide carbon footprint of the computing and telecommunication sectors might triple from 2002 to 2020, the same sectors could deliver over 5 times their footprint in emission savings in other industries (including transportation and energy generation and transmission).[16]

Whether that prediction is accurate depends largely on how smartly computing is deployed in those sectors. It is clear, however, that even if the environmental effect of computing machinery is dwarfed by the environmental savings made possible by its use, computing will remain a large consumer of electricity, so curbing consumption of natural resources should continue to have high priority.

Transitioning of Legacy Applications

It will take time for results of the proposed research agenda to come to fruition. Society has an immediate and pressing need to use current and emerging chip multiprocessor systems effectively. To that end, the committee offers two recommendations related to current development and engineering practices.

Although we expect long-term success in the effective use of parallel systems to come from rethinking architectures and algorithms and developing new programming methods, this strategy will probably sacrifice the backward-platform and cross-platform compatibility that has been an economic cornerstone of IT for decades. To salvage value from the nation's current, substantial IT investment, we should seek ways to bring sequential programs into the parallel world. On the one hand, we expect no silver bullets to enable automatic black-box transformation. On the other hand, it is prohibitively expensive to rewrite many applications. In fact, the committee believes that industry will not migrate its installed base of software to a new parallel future without good, reliable tools to facilitate the migration. Not only can industry not afford a brute-force migration financially, but also it cannot take the chance that innate latent bugs will manifest, potentially many years after the original software engineers created the code being migrated. If we cannot find a way to smooth the transition, this single item could stall the entire parallelism effort, and innovation in many types of IT might well stagnate. The committee urges industry and academe to develop tools that provide a

---

[16]Global e-Sustainability Initiative, 2008, Smart2020: Enabling the Low Carbon Economy in the Information Age, Brussels, Belgium: Global e-Sustainability Initiative, available online at http://www.smart2020.org.

middle ground and give experts "power tools" that can assist with the hard work that will be necessary for vastly increased parallelization. In addition, emphasis should be placed on tools and strategies to enhance code creation, maintenance, verification, and adaptation. All are essential, and current solutions, which are often inadequate even for single-thread software development, are unlikely to be useful for parallel systems.

Recommendation: Invest in the development of tools and methods to transform legacy applications to parallel systems.

Interface Standards

Competition in the private sector often (appropriately) encourages the development of proprietary interfaces and implementations that seek to create competitive advantage. In computer systems, however, a lack of standardization can also impede progress when many incompatible approaches allow none to achieve the benefits of wide adoption and reuse—and this is a major reason that industry participates in standards efforts. We therefore encourage the development of programming interface standards. Standards can facilitate wide adoption of parallel programming and yet encourage competition that will benefit all. Perhaps a useful model is the one used for Java: the standard was initially developed by a small team (not a standards committee), protected in incubation from devolving into many incompatible variants, and yet made public enough to facilitate use and adoption by many cooperating and competing entities.

Recommendation: To promote cooperation and innovation by sharing, encourage development of open interface standards for parallel programming rather than proliferating proprietary programming environments.

PARALLEL-PROGRAMMING MODELS AND EDUCATION

As described earlier in this report, future growth in performance will be driven by parallel programs. Because most programs now in use are not parallel, we will need to rely on the creation of new parallel programs. Who will create those programs? Students must be educated in parallel programming at both the undergraduate and the graduate levels, both in computer science and in other domains in which specialists use computers.

Current State of Programming

One view of the current pool of practicing programmers is that there is a large disparity between the very best programmers and the rest in both time to solution and elegance of solution. The conventional wisdom in the field is that the difference in skills and productivity between the average programmer and the best programmers is a factor of 10 or more.[17] Opinions may vary on the specifics, but the pool of programmers breaks down roughly as follows:

A. A few highly trained, highly skilled, and highly productive computer science (CS) system designers.
B. A few highly trained, highly skilled, and highly productive CS application developers.
C. Many moderately well-trained (average), moderately productive CS system developers.
D. Many moderately productive developers without CS training.

The developers who are not CS-trained are domain scientists, business people, and others who use computers as a tool to solve their problems. There are many such people. It is possible that fewer of those people will be able to program well in the future, because of the difficulty of parallel programming. However, if the CS community develops good abstractions and programming languages that make it easy to program in parallel, even more of those types of developers will be productive.

There is some chance that we will find solutions in which most programmers still program sequentially. Some existing successful systems, such as databases and Web services, exploit parallelism but do not require parallel programs to be written by most users and developers. For example, a developer writes a single-threaded database query that operates in parallel with other queries managed by the database system. Another more modern and popularly known example is MapReduce, which abstracts many programming problems for search and display into a sequence of Map and Reduce operations, as described in Chapter 4.

Those examples are compelling and useful, but we cannot assume that such domain-specific solutions will generalize to all important and pervasive problems. In addition to the shift to new architectural approaches,

---

[17]In reality, the wizard programmers can have an even far greater effect on the organization than the one order of magnitude cited. The wizards will naturally gravitate to an approach to problems that saves tremendous amounts of effort and will debug later, and they will keep a programming team out of trouble far out of proportion to the 10:1 ratio mentioned. Indeed, as in arts in general, there is a sense in which no number of ordinary people can be combined to accomplish what one gifted person can contribute.

attention must be paid to the curriculum to ensure that students are prepared to keep pace with the expected changes in software systems and development. Without adequate training, we will not produce enough of the category (A) and (B) highly skilled programmers above. Without them, who will build the programming abstraction systems?

Parallel computing and thus parallel programming showed great promise in the 1980s with comparably great expectations about what could be accomplished. However, apart from horizontally scalable programming paradigms, such as MapReduce, limited progress resulted in frustration and comparatively little progress in recent years. Accordingly, the focus recently has been more on publishable research results on the theory of parallelism and new languages and approaches and less on simplification of expression and practical use of parallelism and concurrency.

There has been much investment and comparatively limited success in the approach of automatically extracting parallelism from sequential code. There has been considerably less focus on effective expression of parallelism in such a way that software is not expected to guess what parallelism was present in the original problem or computational formulation. Those questions remain unresolved. What should we teach?

Modern Computer-Science Curricula Ill-Equipped for a Parallel Future

In the last 20 years, what is considered CS has greatly expanded, and it has been increasingly interdisciplinary. Recently, many CS departments—such as those at the Massachusetts Institute of Technology, Cornell University, Stanford University, and the Georgia Institute of Technology—have revised their curricula by reducing or eliminating a required core and adding multiple "threads" of concentrations from which students choose one or more specializations, such as computational biology, computer systems, theoretical computing, human-computer interaction, graphics, robotics, or artificial intelligence. With respect to the topic of the present report, the CS curriculum is not training undergraduate and graduate students in either effective parallel programming or parallel computational thinking. But that knowledge is now necessary for effective programming of current commodity-parallel hardware, which is increasingly common in the form of CMPs and graphics processors, not to mention possible changes in systems of the future.

Developers and system designers are needed. Developers design and program application software; system designers design and build parallel-programming systems—which include programming languages, compilers, runtime systems, virtual machines, and operating systems—to make them work on computer hardware. In most universities, parallel programming is not part of the undergraduate curriculum for either CS students or scientists in other domains and is often presented as a graduate elective course for CS and electrical and computer engineering students. In the coming world of nearly ubiquitous parallel architectures, relegating parallelism to the boundaries of the curriculum will not suffice. Instead, it will increasingly be a practical tool for domain scientists and will be immediately useful for software, system, and application development.

Parallel programming—even the parallel programming of today—is hard, but there are enough counterexamples to suggest that it may not be intractable. Computational reasoning for parallel problem-solving—the intellectual process of mapping the structure of a problem to a strategy for solution—is fairly straightforward for computer scientists and domain scientists alike, regardless of the level of parallelism involved or apparent in the solution. Most domain scientists—in such fields as physics, biology, chemistry, and engineering—understand the concepts of causality, correlation, and independence (parallelism vs sequence). There is a mismatch between how scientists and other domain specialists think about their problems and how they express parallelism in their code. It therefore becomes difficult for both computer and noncomputer scientists to write programs. Straightforwardness is lost in the current expression of parallel programming. It is possible, and even common, to express the problem of parallel programming in a way that is complex and difficult to understand, but the recommendations in this report are aimed at developing models and approaches in which such complexity is not necessary.

Arguably, computational experimentation—performing science exploration with computer models—is becoming an important part of modern scientific endeavor. Computational experimentation is modernizing the scientific method. Consequently, the ability to express scientific theories and models in computational form is a critical skill for modern scientists. If computational models are to be targeted to parallel hardware, as we argue in this report, parallel approaches to reasoning and thinking will be essential. Jeannette Wing has argued[18] for the importance of computational thinking, broadly, and a current National Research Council study is exploring that notion. A recent report of that study also touched on concurrency and parallelism as part of computational thinking.[19] With respect to the CS curriculum, because no general-purpose paradigm has

---

[18]Jeannette M. Wing, 2006, Computational thinking, Communications of the ACM 49(3): 33-35.

[19]See NRC, 2010, Report of a Workshop on the Scope and Nature of Computational Thinking, Washington, D.C.: The National Academies Press, available online at http://www.nap.edu/catalog.php?record_id=12840.

emerged, universities should teach diverse parallel-programming languages, abstractions, and approaches until effective ways of teaching and programming emerge. The necessary shape of the needed changes will not be clear until some reasonably general parallel-programming methods have been devised and shown to be promising. Nevertheless, possible models for reform include making parallelism an intrinsic part of every course (algorithms, architecture, programming, operating systems, compilers, and so on) as a fundamental way of solving problems; adding specialized courses, such as parallel computational reasoning, parallel algorithms, parallel architecture, and parallel programming; and creating an honors section for advanced training in parallelism (this option is much less desirable in that it enforces the notion that parallel programming is outside mainstream approaches). It will be important to try many parallel-programming languages and models in the curriculum and in research to sort out which ones will work best and to learn the most effective methods.

Recommendation: Incorporate in computer science education an increased emphasis on parallelism, and use a variety of methods and approaches to prepare students better for the types of computing resources that they will encounter in their careers.

GAME OVER OR NEXT LEVEL?

Since the invention of the transistor and the stored-program computer architecture in the 1ate 1940s, we have enjoyed over a half-century of phenomenal growth in computing and its effects on society. Will the second half of the 20th century be recorded as the golden age of computing progress, or will we now step up to the next set of challenges and continue the growth in computing that we have come to expect?

Our computing models are likely to continue to evolve quickly in the foreseeable future. We expect that there are still many changes to come, which will require evolution of combined software and hardware systems. We are already seeing substantial centralization of computational capability in the cloud-computing paradigm with its attendant challenges to data storage and bandwidth. It is also possible to envision an abundance of Internet-enabled embedded devices that run software that has the sophistication and complexity of software running on today's general-purpose processors. Networked, those devices will form a ubiquitous and invisible computing platform that provides data and services that we can only begin to imagine today. These drivers combine with the technical constraints and challenges outlined in the rest of this report to reinforce the notion that computing is changing at virtually every level.

The end of the exponential runup in uniprocessor performance and the market saturation of the general-purpose processor mark the end of the "killer micro." This is a golden time for innovation in computing architectures and software. We have already begun to see diversity in computer designs to optimize for such metrics as power and throughput. The next generation of discoveries will require advances at both the hardware and the software levels.

There is no guarantee that we can make future parallel computing ubiquitous and as easy to use as yesterday's sequential computer, but unless we aggressively pursue efforts suggested by the recommendations above, it will be game over for future growth in computing performance. This report describes the factors that have led to the limitations on growth in the use of single processors based on CMOS technology. The recommendations here are aimed at supporting and focusing research, development, and education in architectures, power, and parallel computing to sustain growth in computer performance and enjoy the next level of benefits to society.

Appendixes

A

A History of Computer Performance

Computer performance has historically been defined by how fast a computer system can execute a single-threaded program to perform useful work. Why care about computer performance? What is the work? How has computer performance improved?

Better computer performance matters in two ways. First, in what is often called capability computing, it can enable computations that were previously not practical or worthwhile. It does no good to compute tomorrow's weather forecast in 24 hours, but 12-hour computation is valuable. Second, when performance scales up more rapidly than computer cost—as has often been the case—better cost performance allows computation to be used where it was previously not economically tenable. Neither spreadsheets on $1,000,000 mainframes nor $10,000 MP3 players make sense.

Computer performance should be evaluated on the basis of the work that matters. Computer vendors should analyze their designs with the (present and future) workloads of their (present and future) customers, and those purchasing computers should consider their own (present and future) workloads with alternative computers under consideration. Because the above is time-consuming—and therefore expensive—many people evaluate computers by using standard benchmark suites. Each vendor produces benchmark results for its computers, often after optimizing computers for the benchmarks. Each customer can then compare benchmark results and get useful information—but only if the benchmark suite is sufficiently close to the customer's actual workloads.

Two popular benchmark suites are SPECint2000 and SPECfp2000. Both are produced by the Standard Performance Evaluation Corporation (SPEC) (http://www.spec.org/). SPECint2000 includes 12 integer codes, and SPECfp200 has 14 floating-point benchmarks. Below, we use SPEC data to examine computer-performance trends over the last 2 decades. The results are valuable, but their absolute numbers should be regarded as rough approximations of systems' absolute performance. Nevertheless, they are much better than results based on "peak rate," which gives a computer's speed when it is doing nothing.

Figures A.1 (INT) and A.2 (FP) display results for SPECint2000 and SPECfp2000, respectively. The X axes give the years from 1985 or 1988 to 2007. The logarithmic Y axes give the SPEC rate normalized to circa 1985. Thus, a value of 10 means that the computer is 10 times faster than (can execute the work in one-tenth the time of) a 1985 model.

The Figures A.1 and A.2 reveal two trends. First, computer performance has improved exponentially (linearly on a semilogarithmic plot) for most years under study. In particular, until 2004 or so, both SPECint2000 and SPECfp2000 improved at a compound annual rate exceeding 50% (for example, a factor of 100 in about 10 years).

Second, the performance improvements after 2004 have been poorer.

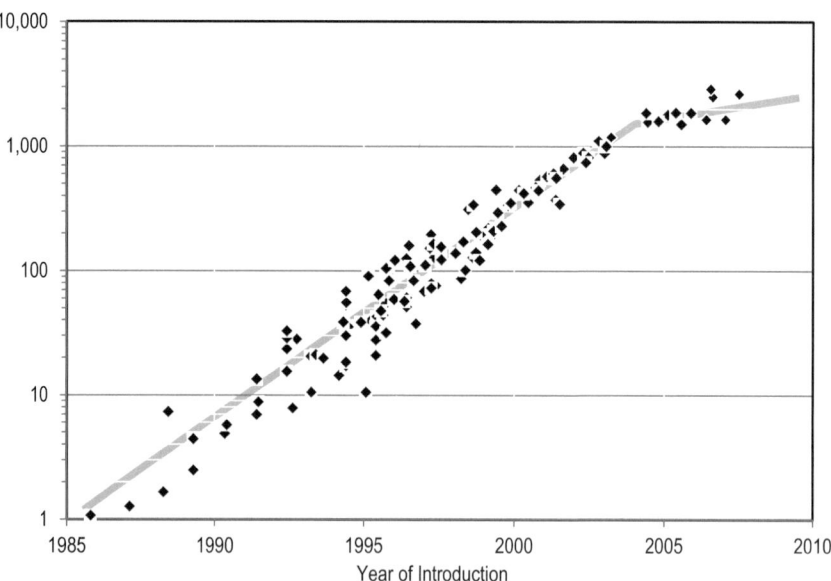

FIGURE A.1 Integer application performance (SPECint2000) over time (1985-2010).

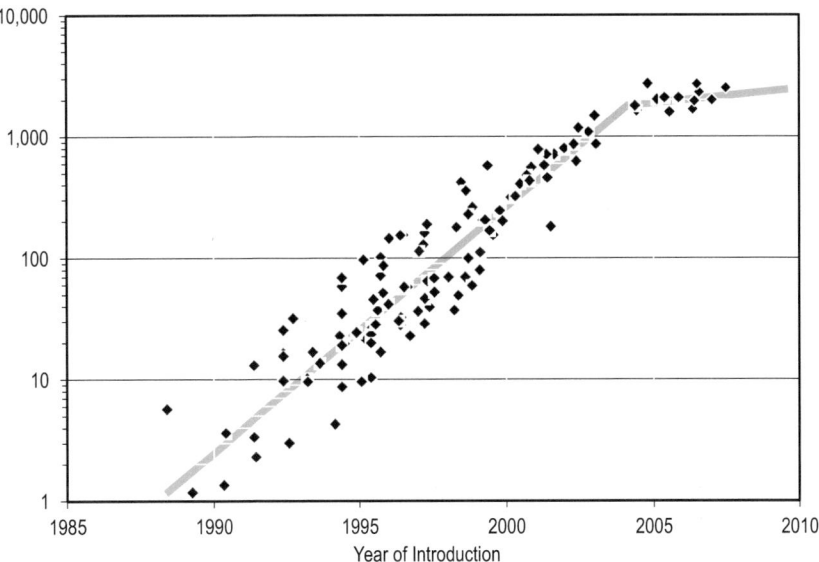

FIGURE A.2 Floating-point application performance (SPECfp2000) over time (1985-2010).

One could hope that the results are an anomaly and that computer vendors will soon return to robust annual improvements. However, public roadmaps and private conversations with vendors reveal that single-threaded computer-performance gains have entered a new era of modest improvement.

Trends in computer-clock frequency offer another reason for pessimism. Clock frequency is the "heart rate" of a computer, and its improvement has traditionally been a major component of computer-performance improvement. Figure A.3 (FREQ) illustrates clock frequency over time in megahertz (millions of cycles per second). Clearly, clock-frequency improvements have also stalled (especially if the 4.7-GHz power 6 is more an exception than the new rule).

Moreover, the principal reason that clock-frequency improvement has slowed greatly is that higher clock frequencies demand greater power and the power used by modern microprocessors has reached a level that make increases questionable from an economic perspective and may even encourage clock-frequency reductions. Figure A.4 (POWER) plots chip power (in watts) versus. year. Like clock frequencies, power consumed by a chip increased exponentially (linearly on a semilogarithmic plot) for years, but it has recently reached a plateau.

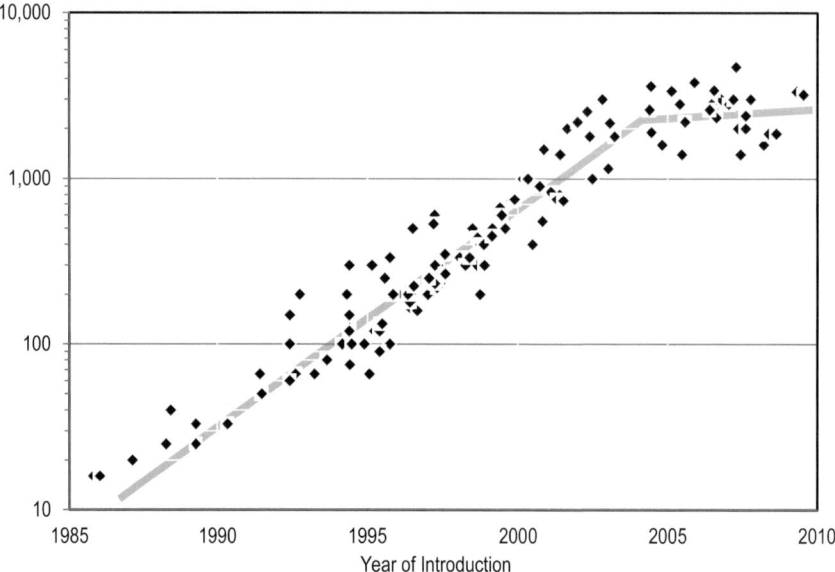

FIGURE A.3 Microprocessor clock frequency (MHz) over time (1985-2010).

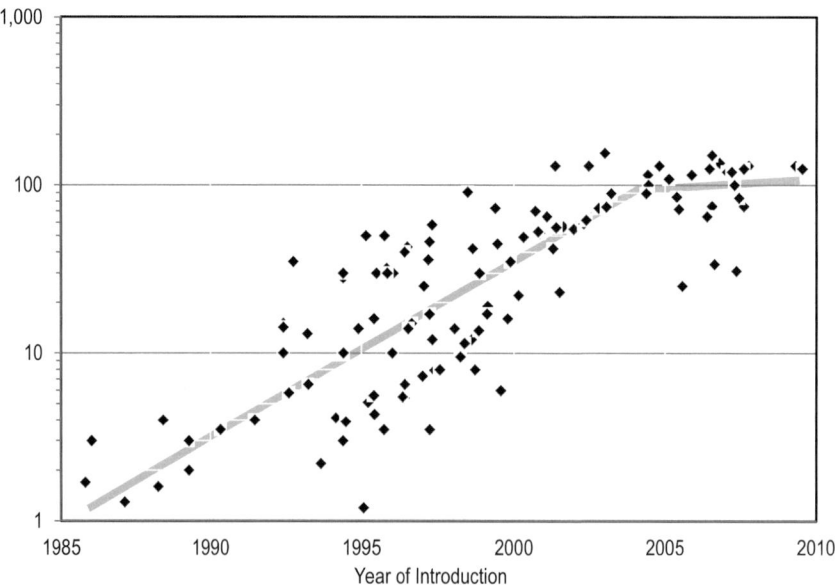

FIGURE A.4 Microprocessor power dissipation (watts) over time (1985-2010).

To put those trends into context, we should look further back in history. Kurzweil and others argue that the performance improvements during the recent microprocessor or Moore's law era follow a longer historical trend.[1] In particular, in Kurzweil's estimates of 20th-century computer-performance improvements, he finds that over the whole century, computer performance improved exponentially and that exponential improvement was, if anything, accelerating. Although his exact numbers are open to debate, it is clear that computer performance grew exponentially over the entirety of the 20th century.

What will the rest of the 21st century bring? Optimists predict that Kurzweil's trend will continue. Pessimists worry that single-threaded architectures and complementary metal oxide semiconductor technology are reaching their limits, that multithreaded programming has not been broadly successful, and that alternative technologies are still insufficient. Our job, as encouraged in the rest of this report, is to prove the optimists correct.

---

[1] The Law of Accelerating Returns, by Ray Kurzweil, http://www.kurzweilai.net/articles/art0134.html?printable=1.

B

Biographies of Committee Members and Staff

Samuel H. Fuller (Chair), NAE, is the CTO and vice president of research and development at Analog Devices, Inc. (ADI) and is responsible for its technology and product strategy. He also manages university research programs and advanced development initiatives and supports the growth of ADI product-design centers around the world. Dr. Fuller has managed the development of EDA tools and methods and design of digital signal processors and sponsored the development of advanced optoelectronic integrated circuits. Before joining ADI in 1998, Dr. Fuller was vice president of research at Digital Equipment Corporation and built the company's corporate research programs, which included laboratories in Massachusetts, California, France, and Germany. While at Digital, he initiated work in local-area networking, RISC processors, distributed systems, and Internet search engines. He was also responsible for research programs with universities; the Massachusetts Institute of Technology Project Athena was one of the major programs. Earlier, Dr. Fuller was an associate professor of computer science and electrical engineering at Carnegie Mellon University, where he led the design and performance evaluation of experimental multiprocessor computer systems. He holds a BS in electrical engineering from the University of Michigan and an MS and a PhD from Stanford University. He is a member of the board of Zygo Corporation and the Corporation for National Research Initiatives and serves on the Technology Strategy Committee of the Semiconductor Industry Association. Dr. Fuller has served on several National Research Council studies, including the one that produced *Cryptography's*

Role in Securing the Information Society, and was a founding member of the Research Council's Computer Science and Telecommunications Board. He is a fellow of the Institute of Electrical and Electronics Engineers and the American Association for the Advancement of Science and a member of the National Academy of Engineering.

Luiz André Barroso is a Distinguished Engineer at Google Inc., where his work has spanned a number of fields, including software-infrastructure design, fault detection and recovery, power provisioning, networking software, performance optimizations, and the design of Google's computing platform. Before joining Google, he was a member of the research staff at Compaq and Digital Equipment Corporation, where his group did some of the pioneering work on processor and memory-system design for commercial workloads (such as database and Web servers). The group also designed Piranha, a scalable shared-memory architecture based on single-chip multiprocessing; this work on Piranha has had an important impact in the microprocessor industry, helping to inspire many of the multicore central processing units that are now in the mainstream. Before joining Digital, he was one of the designers of the USC RPM, an FPGA-based multiprocessor emulator for rapid hardware prototyping. He has also worked at IBM Brazil's Rio Scientific Center and lectured at PUC-Rio (Brazil) and Stanford University. He holds a PhD in computer engineering from the University of Southern California and a BS and an MS in electrical engineering from the Pontifícia Universidade Católica, Rio de Janeiro.

Robert P. Colwell, NAE, was Intel's chief IA32 (Pentium) microprocessor architect from 1992 to 2000 and managed the IA32 Architecture group at Intel's Hillsboro, Oregon, facility through the P6 and Pentium 4 projects. He was named the Eckert-Mauchly Award winner for 2005. He was elected to the National Academy of Engineering in 2006 "for contributions to turning novel computer architecture concepts into viable, cutting-edge commercial processors." He was named an Intel fellow in 1996, and a fellow of the Institute of Electrical and Electronics Engineers (IEEE) in 2006. Previously, Dr. Colwell was a central processing unit architect at VLIW minisupercomputer pioneer Multiflow Computer, a hardware-design engineer at workstation vendor Perq Systems, and a member of technical staff at Bell Labs. He has published many technical papers and journal articles, is inventor or coinventor on 40 patents, and has participated in numerous panel sessions and invited talks. He is the Perspectives editor for IEEE's *Computer* magazine, wrote the At Random column in 2002-2005, and is author of *The Pentium Chronicles*, a behind-the-scenes look at modern microprocessor design. He is currently an independent consultant. Dr.

Colwell holds a BSEE from the University of Pittsburgh and an MSEE and a PhD from Carnegie Mellon University.

William J. Dally, NAE, is the Willard R. and Inez Kerr Bell Professor of Engineering at Stanford University and chair of the Computer Science Department. He is also chief scientist and vice president of NVIDIA Research. He has done pioneering development work at Bell Telephone Laboratories, the California Institute of Technology, and the Massachusetts Institute of Technology, where he was a professor of electrical engineering and computer science. At Stanford University, his group has developed the Imagine processor, which introduced the concepts of stream processing and partitioned register organizations. Dr. Dally has worked with Cray Research and Intel to incorporate many of those innovations into commercial parallel computers and with Avici Systems to incorporate the technology into Internet routers, and he cofounded Velio Communications to commercialize high-speed signaling technology and Stream Processors to commercialize stream-processor technology. He is a fellow of the Institute of Electrical and Electronics Engineers and of the Association for Computing Machinery (ACM) and has received numerous honors, including the ACM Maurice Wilkes award. He has published more than 150 papers and is an author of the textbooks *Digital Systems Engineering* (Cambridge University Press, 1998) and *Principles and Practices of Interconnection Networks* (Morgan Kaufmann, 2003). Dr. Dally is a member of the Computer Science and Telecommunications Board (CSTB) and was a member of the CSTB committee that produced the report *Getting up to Speed: The Future of Supercomputing*.

Dan Dobberpuhl, NAE, cofounder, president, and CEO of P. A. Semi, has been credited with developing fundamental breakthroughs in the evolution of high-speed and low-power microprocessors. Before starting P. A. Semi, Mr. Dobberpuhl was vice president and general manager of the broadband processor division of Broadcom Corporation. He came to Broadcom via an acquisition of his previous company, SiByte, Inc., founded in 1998, which was sold to Broadcom in 2000. Before that, he worked for Digital Equipment Corporation for more than 20 years, where he was credited with creating some of the most fundamental breakthroughs in microprocessing technology. In 1998, *EE Times* named Mr. Dobberpuhl as one of the "40 forces to shape the future of the Semiconductor Industry." In 2003, he was awarded the prestigious IEEE Solid State Circuits Award for "pioneering design of high-speed and low-power microprocessors." In 2006, Mr. Dobberpuhl was elected to the National Academy of Engineering for "innovative design and implementation of high-performance, low-power microprocessors." Mr. Dobberpuhl holds

15 patents and has written many publications related to integrated circuits and central processing units, including being coauthor of the seminal textbook *Design and Analysis of VLSI Circuits*, published by Addison-Wesley in 1985. He holds a bachelor's degree in electrical engineering from the University of Illinois.

Pradeep Dubey is a senior principal engineer and director of the Parallel Computing Lab, part of Intel Labs at Intel Corporation. His research focus is computer architectures to handle new application paradigms for the future computing environment efficiently. Dr. Dubey previously worked at IBM's T. J. Watson Research Center and Broadcom Corporation. He was one of the principal architects of the AltiVec* multimedia extension to Power PC* architecture. He also worked on the design, architecture, and performance issues of various microprocessors, including Intel® i386™, i486™, and Pentium® processors. He holds 26 patents and has published extensively. Dr. Dubey received a BS in electronics and communication engineering from Birla Institute of Technology, India, an MSEE from the University of Massachusetts at Amherst, and a PhD in electrical engineering from Purdue University. He is a fellow of the Institute of Electrical and Electronics Engineering.

Mark D. Hill is a professor in both the Computer Sciences Department and the Electrical and Computer Engineering Department at the University of Wisconsin-Madison. Dr. Hill's research targets the memory systems of multiple-processor and single-processor computer systems. His work emphasizes quantitative analysis of system-level performance. His research interests include parallel computer-system design (for example, memory-consistency models and cache coherence), memory-system design (for example, caches and translation buffers), computer simulation (for example, parallel systems and memory systems), and software (page tables and cache-conscious optimizations for databases and pointer-based codes). He is the inventor of the widely used 3C model of cache behavior (compulsory, capacity, and conflict misses). Dr. Hill's current research is mostly part of the Wisconsin Multifacet Project that seeks to improve the multiprocessor servers that form the computational infrastructure for Internet Web servers, databases, and other demanding applications. His work focuses on using the transistor bounty predicted by Moore's law to improve multiprocessor performance, cost, and fault tolerance while making these systems easier to design and program. Dr. Hill is a fellow of the Association for Computing Machinery (ACM) (2004) for contributions to memory-consistency models and memory-system design and a fellow of the Institute of Electrical and Electronics Engineers (2000) for contributions to cache-memory design and analysis. He was named a

Wisconsin Vilas Associate in 2006, was a co-winner of the best-paper award in VLDB in 2001, was named a Wisconsin Romnes fellow in 1997, and won a National Science Foundation Presidential Young Investigator award in 1989. He is a director of ACM SIGARCH, coeditor of *Readings in Computer Architecture* (2000), and coinventor on 28 U.S. patents (several coissued in the European Union and Japan). He has held visiting positions at Universidad Politecnica de Catalunya (2002-2003) and Sun Microsystems (1995-1996). Dr. Hill earned a PhD in computer science from the University of California, Berkeley (UCB) in 1987, an MS in computer science from UCB in 1983, and a BSE in computer engineering from the University of Michigan-Ann Arbor in 1981.

Mark Horowitz, NAE, is the associate vice provost for graduate education, working on special programs, and the Yahoo! Founders Professor of the School of Engineering at Stanford University. In addition, he is chief scientist at Rambus Inc. He received his BS and MS in electrical engineering from the Massachusetts Institute of Technology in 1978 and his PhD from Stanford in 1984. Dr. Horowitz has received many awards, including a 1985 Presidential Young Investigator Award, the 1993 ISSCC Best Paper Award, the ISCA 2004 Most Influential Paper of 1989, and the 2006 Don Pederson IEEE Technical Field Award. He is a fellow of the Institute of Electrical and Electronics Engineers and the Association for Computing Machinery and is a member of the National Academy of Engineering. Dr. Horowitz's research interests are quite broad and span using electrical engineering and computer science analysis methods on problems in molecular biology and creating new design methods for analogue and digital very-large-scale implementation circuits. He has worked on many processor designs, from early RISC chips to creating some of the first distributed shared-memory multiprocessors and is currently working on on-chip multiprocessor designs. Recently, he has worked on a number of problems in computational photography. In 1990, he took leave from Stanford to help start Rambus Inc., a company designing high-bandwidth memory-interface technology, and has continued work in high-speed I/O at Stanford. His current research includes multiprocessor design, low-power circuits, high-speed links, computational photography, and applying engineering to biology.

David Kirk, NAE, was NVIDIA's chief scientist since from 1997 to 2009 and is now an NVIDIA fellow. His contributions include leading NVIDIA graphics-technology development for today's most popular consumer entertainment platforms. In 2002, Dr. Kirk received the SIGGRAPH Computer Graphics Achievement Award for his role in bringing high-performance computer graphics systems to the mass market. From 1993 to 1996, he was chief scientist and head of technology for Crystal Dynamics, a video-game manufacturing company. From 1989 to 1991, Dr. Kirk was an engineer for the Apollo Systems Division of Hewlett-Packard Company. He is the inventor on 50 patents and patent applications related to graphics design and has published more than 50 articles on graphics technology. Dr. Kirk holds a BS and an MS in mechanical engineering from the Massachusetts Institute of Technology and an MS and a PhD in computer science from the California Institute of Technology.

Monica Lam is a professor of computer science at Stanford University, having joined the faculty in 1988. She has contributed to research on a wide array of computer-systems topics, including compilers, program analysis, operating systems, security, computer architecture, and high-performance computing. Her recent research focus is to make computing and programming easier. In the Collective Project, she and her research group developed the concept of a livePC: subscribers to the livePC will automatically run the latest of the published PC virtual images with each reboot. That approach allows computers to be managed scalably and securely. In 2005, the group started a company called moka5 to transfer the technology to industry. In another research project, her program-analysis group has developed a collection of tools for improving software security and reliability. They developed the first scalable context-sensitive inclusion-based pointer analysis and a freely available tool called BDDBDDB that allows programmers to express context-sensitive analyses simply by writing Datalog queries. Other tools developed include Griffin, static and dynamic analysis for finding security vulnerabilities in Web applications, such as SQL injection; a static and dynamic program query language called PQL; a static memory-leak detector called Clouseau; a dynamic buffer-overrun detector called CRED; and a dynamic error-diagnosis tool called DIDUCE. Previously, Dr. Lam led the Stanford University Intermediate Format Compiler project, which produced a widely used compiler infrastructure known for its locality optimizations and interprocedural parallelization. Many of the compiler techniques that she developed have been adopted by industry. Her other research projects included the architecture and compiler for the CMU Warp machine, a systolic array of very-long-instruction-word processors, and the Stanford DASH distributed shared-memory machine. In 1998, she took a sabbatical leave from Stanford University to help to start Tensilica Inc., a company that specializes in configurable processor cores. She received a BSc from the University of British Columbia in 1980 and a PhD in computer science from Carnegie Mellon University in 1987.

Kathryn S. McKinley is a professor at the University of Texas at Austin. Her research interests include compilers, runtime systems, and architecture. Her research seeks to enable high-level programming languages to achieve high performance, reliability, and availability. She and her collaborators have developed compiler optimizations for improving memory-system performance, high-performance garbage-collection algorithms, scalable explicit-memory management algorithms for parallel systems, and cooperative dynamic optimizations for improving the performance of managed languages. She is leading the compiler effort for the TRIPS project, which is exploring attaining scalable performance improvements using explicit dataflow graph execution architectures. Her honors include being named an Association for Computing Machinery (ACM) Distinguished Scientist and receiving a National Science Foundation Career Award. She is the co-editor-in-chief of ACM's *Transactions on Programming Language and Systems (TOPLAS)*. She is active in increasing minority-group participation in computer science and, for example, co-led with Daniel Jimenez the CRAW/CDC Programming Languages Summer School in 2007. She has published over 75 refereed articles and has supervised eight PhD degrees. Dr. McKinley holds a BA (1985) in electrical engineering and computer science and an MS (1990) and a PhD (1992) in computer science, all from Rice University.

Charles Moore is an Advanced Micro Devices (AMD) corporate fellow and the CTO for AMD's Technology Group. He is the chief engineer of AMD's next-generation processor design. His responsibilities include interacting with key customers to understand their requirements, identifying important technology trends that may affect future designs, and architectural development and management of the next-generation design. Before joining AMD, Mr. Moore was a senior industrial research fellow at the University of Texas at Austin, where he did research on technology-scalable computer architecture. Before then, he was a distinguished engineer at IBM, where he was the chief engineer on the POWER4 project. Earlier, he was the coleader of the first single-chip POWER architecture implementation and the coleader of the first PowerPC implementation used by Apple Computer in its PowerMac line of personal computers. While at IBM, he was elected to the IBM Academy of Technology and was named an IBM master inventor. He has been granted 29 US patents and has several others pending. He has published numerous conference papers and articles on a wide array of subjects related to computer architecture and design. He is on the editorial board of *IEEE Micro* magazine and on the program committee for several important industry conferences. Mr. Moore holds a master's degree in electrical engineering from the University of Texas at

Austin and a bachelor's degree in electrical engineering from the Rensselaer Polytechnic Institute.

Katherine Yelick is a professor in the Computer Science Division of the University of California, Berkeley. The main goal of her research is to develop techniques for obtaining high performance on a wide array of computational platforms and to ease the programming effort required to obtain performance. Dr. Yelick is perhaps best known for her efforts in global address space (GAS) languages, which attempt to present the programmer with a shared-memory model for parallel programming. Those efforts have led to the design of Unified Parallel C (UPC), which merged some of the ideas of three shared-address-space dialects of C: Split-C, AC (from IDA), and PCP (from Lawrence Livermore National Laboratory). In recent years, UPC has gained recognition as an alternative to message-passing programming for large-scale machines. Compaq, Sun, Cray, HP, and SGI are implementing UPC, and she is currently leading a large effort at Lawrence Berkeley National Laboratory to implement UPC on Linux clusters and IBM machines and to develop new optimizations. The language provides a uniform programming model for both shared and distributed memory hardware. She has also worked on other global-address-space languages, such as Titanium, which is based on Java. She has done notable work on single-processor optimizations, including techniques for automatically optimizing sparse matrix algorithms for memory hierarchies. Another field that she has worked in is architectures for memory-intensive applications and in particular the use of mixed logic, which avoids the off-chip accesses to DRAM, thereby gaining bandwidth while lowering latency and energy consumption. In the IRAM project, a joint effort with David Patterson, she developed an architecture to take advantage of this technology. The IRAM processor is a single-chip system designed for low power and high performance in multimedia applications and achieves an estimated 6.4 gigaops per second in a 2-W design. Dr. Yelick received her bachelor's degree (1985), master's degree (1985), and PhD (1991) in electrical engineering and computer science from the Massachusetts Institute of Technology.

STAFF

Lynette I. Millett is a senior program officer and study director at the Computer Science and Telecommunications Board (CSTB), National Research Council of the National Academies. She currently directs several CSTB projects, including a study to advise the Centers for Medicare and Medicaid Service on future information systems architectures and a study examining opportunities for computing research to help meet sustainability challenges. She served as the study director for the CSTB report *Social Security Administration Electronic Service Provision: A Strategic Assessment*. Ms. Millett's portfolio includes substantial portions of CSTB's recent work on software, identity systems, and privacy. She directed, among other projects, those that produced *Software for Dependable Systems: Sufficient Evidence?*, an exploration of fundamental approaches to developing dependable mission-critical systems; *Biometric Recognition: Challenges and Opportunities*, a comprehensive assessment of biometric technology; *Who Goes There? Authentication Through the Lens of Privacy*, a discussion of authentication technologies and their privacy implications; and *IDs—Not That Easy: Questions About Nationwide Identity Systems*, a post-9/11 analysis of the challenges presented by large-scale identity systems. She has an M.Sc. in computer science from Cornell University, where her work was supported by graduate fellowships from the National Science Foundation and the Intel Corporation; and a BA with honors in mathematics and computer science from Colby College, where she was elected to Phi Beta Kappa.

Shenae Bradley is a senior program assistant at the Computer Science and Telecommunications Board of the National Research Council. She currently provides support for the Committee on Sustaining Growth in Computing Performance, the Committee on Wireless Technology Prospects and Policy Options, and the Computational Thinking for Everyone: A Workshop Series Planning Committee, to name a few. Prior to this, she served as an administrative assistant for the Ironworker Management Progressive Action Cooperative Trust and managed a number of apartment rental communities for Edgewood Management Corporation in the Maryland/DC/Delaware metropolitan areas. Ms. Bradley is in the process of earning her BS in family studies from the University of Maryland at College Park.

C

Reprint of Gordon E. Moore's "Cramming More Components onto Integrated Circuits"

NOTE: Reprinted from Gordon Moore, 1965, Cramming more components onto integrated circuits, Electronics 38(8) with permission from Intel Corporation.

Cramming More Components onto Integrated Circuits

GORDON E. MOORE, LIFE FELLOW, IEEE

With unit cost falling as the number of components per circuit rises, by 1975 economics may dictate squeezing as many as 65 000 components on a single silicon chip.

The future of integrated electronics is the future of electronics itself. The advantages of integration will bring about a proliferation of electronics, pushing this science into many new areas.

Integrated circuits will lead to such wonders as home computers—or at least terminals connected to a central computer—automatic controls for automobiles, and personal portable communications equipment. The electronic wristwatch needs only a display to be feasible today.

But the biggest potential lies in the production of large systems. In telephone communications, integrated circuits in digital filters will separate channels on multiplex equipment. Integrated circuits will also switch telephone circuits and perform data processing.

Computers will be more powerful, and will be organized in completely different ways. For example, memories built of integrated electronics may be distributed throughout the machine instead of being concentrated in a central unit. In addition, the improved reliability made possible by integrated circuits will allow the construction of larger processing units. Machines similar to those in existence today will be built at lower costs and with faster turnaround.

I. PRESENT AND FUTURE

By integrated electronics, I mean all the various technologies which are referred to as microelectronics today as well as any additional ones that result in electronics functions supplied to the user as irreducible units. These technologies were first investigated in the late 1950's. The object was to miniaturize electronics equipment to include increasingly complex electronic functions in limited space with minimum weight. Several approaches evolved, including microassembly techniques for individual components, thin-film structures, and semiconductor integrated circuits.

Reprinted from Gordon E. Moore, "Cramming More Components onto Integrated Circuits," *Electronics*, pp. 114–117, April 19, 1965.
Publisher Item Identifier S 0018-9219(98)00753-1.

Each approach evolved rapidly and converged so that each borrowed techniques from another. Many researchers believe the way of the future to be a combination of the various approaches.

The advocates of semiconductor integrated circuitry are already using the improved characteristics of thin-film resistors by applying such films directly to an active semiconductor substrate. Those advocating a technology based upon films are developing sophisticated techniques for the attachment of active semiconductor devices to the passive film arrays.

Both approaches have worked well and are being used in equipment today.

II. THE ESTABLISHMENT

Integrated electronics is established today. Its techniques are almost mandatory for new military systems, since the reliability, size, and weight required by some of them is achievable only with integration. Such programs as Apollo, for manned moon flight, have demonstrated the reliability of integrated electronics by showing that complete circuit functions are as free from failure as the best individual transistors.

Most companies in the commercial computer field have machines in design or in early production employing integrated electronics. These machines cost less and perform better than those which use "conventional" electronics.

Instruments of various sorts, especially the rapidly increasing numbers employing digital techniques, are starting to use integration because it cuts costs of both manufacture and design.

The use of linear integrated circuitry is still restricted primarily to the military. Such integrated functions are expensive and not available in the variety required to satisfy a major fraction of linear electronics. But the first applications are beginning to appear in commercial electronics, particularly in equipment which needs low-frequency amplifiers of small size.

III. RELIABILITY COUNTS

In almost every case, integrated electronics has demonstrated high reliability. Even at the present level of pro-

APPENDIX C

duction—low compared to that of discrete components—it offers reduced systems cost, and in many systems improved performance has been realized.

Integrated electronics will make electronic techniques more generally available throughout all of society, performing many functions that presently are done inadequately by other techniques or not done at all. The principal advantages will be lower costs and greatly simplified design—payoffs from a ready supply of low-cost functional packages.

For most applications, semiconductor integrated circuits will predominate. Semiconductor devices are the only reasonable candidates presently in existence for the active elements of integrated circuits. Passive semiconductor elements look attractive too, because of their potential for low cost and high reliability, but they can be used only if precision is not a prime requisite.

Silicon is likely to remain the basic material, although others will be of use in specific applications. For example, gallium arsenide will be important in integrated microwave functions. But silicon will predominate at lower frequencies because of the technology which has already evolved around it and its oxide, and because it is an abundant and relatively inexpensive starting material.

IV. Costs and Curves

Reduced cost is one of the big attractions of integrated electronics, and the cost advantage continues to increase as the technology evolves toward the production of larger and larger circuit functions on a single semiconductor substrate. For simple circuits, the cost per component is nearly inversely proportional to the number of components, the result of the equivalent piece of semiconductor in the equivalent package containing more components. But as components are added, decreased yields more than compensate for the increased complexity, tending to raise the cost per component. Thus there is a minimum cost at any given time in the evolution of the technology. At present, it is reached when 50 components are used per circuit. But the minimum is rising rapidly while the entire cost curve is falling (see graph). If we look ahead five years, a plot of costs suggests that the minimum cost per component might be expected in circuits with about 1000 components per circuit (providing such circuit functions can be produced in moderate quantities). In 1970, the manufacturing cost per component can be expected to be only a tenth of the present cost.

The complexity for minimum component costs has increased at a rate of roughly a factor of two per year (see graph). Certainly over the short term this rate can be expected to continue, if not to increase. Over the longer term, the rate of increase is a bit more uncertain, although there is no reason to believe it will not remain nearly constant for at least ten years. That means by 1975, the number of components per integrated circuit for minimum cost will be 65 000.

I believe that such a large circuit can be built on a single wafer.

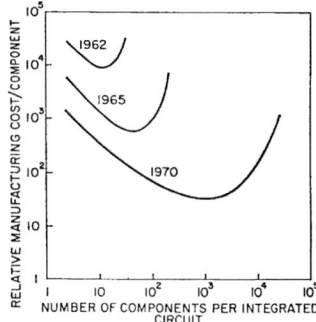

Fig. 1.

V. Two-Mil Squares

With the dimensional tolerances already being employed in integrated circuits, isolated high-performance transistors can be built on centers two-thousandths of an inch apart. Such a two-mil square can also contain several kilohms of resistance or a few diodes. This allows at least 500 components per linear inch or a quarter million per square inch. Thus, 65 000 components need occupy only about one-fourth a square inch.

On the silicon wafer currently used, usually an inch or more in diameter, there is ample room for such a structure if the components can be closely packed with no space wasted for interconnection patterns. This is realistic, since efforts to achieve a level of complexity above the presently available integrated circuits are already under way using multilayer metallization patterns separated by dielectric films. Such a density of components can be achieved by present optical techniques and does not require the more exotic techniques, such as electron beam operations, which are being studied to make even smaller structures.

VI. Increasing the Yield

There is no fundamental obstacle to achieving device yields of 100%. At present, packaging costs so far exceed the cost of the semiconductor structure itself that there is no incentive to improve yields, but they can be raised as high as is economically justified. No barrier exists comparable to the thermodynamic equilibrium considerations that often limit yields in chemical reactions; it is not even necessary to do any fundamental research or to replace present processes. Only the engineering effort is needed.

In the early days of integrated circuitry, when yields were extremely low, there was such incentive. Today ordinary integrated circuits are made with yields comparable with those obtained for individual semiconductor devices. The same pattern will make larger arrays economical, if other considerations make such arrays desirable.

Fig. 2.

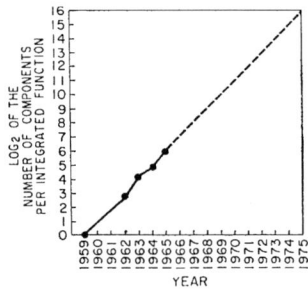

Fig. 3.

VII. Heat Problem

Will it be possible to remove the heat generated by tens of thousands of components in a single silicon chip?

If we could shrink the volume of a standard high-speed digital computer to that required for the components themselves, we would expect it to glow brightly with present power dissipation. But it won't happen with integrated circuits. Since integrated electronic structures are two dimensional, they have a surface available for cooling close to each center of heat generation. In addition, power is needed primarily to drive the various lines and capacitances associated with the system. As long as a function is confined to a small area on a wafer, the amount of capacitance which must be driven is distinctly limited. In fact, shrinking dimensions on an integrated structure makes it possible to operate the structure at higher speed for the same power per unit area.

VIII. Day of Reckoning

Clearly, we will be able to build such component-crammed equipment. Next, we ask under what circumstances we should do it. The total cost of making a particular system function must be minimized. To do so, we could amortize the engineering over several identical items, or evolve flexible techniques for the engineering of large functions so that no disproportionate expense need be borne by a particular array. Perhaps newly devised design automation procedures could translate from logic diagram to technological realization without any special engineering.

It may prove to be more economical to build large systems out of smaller functions, which are separately packaged and interconnected. The availability of large functions, combined with functional design and construction, should allow the manufacturer of large systems to design and construct a considerable variety of equipment both rapidly and economically.

IX. Linear Circuitry

Integration will not change linear systems as radically as digital systems. Still, a considerable degree of integration will be achieved with linear circuits. The lack of large-value capacitors and inductors is the greatest fundamental limitation to integrated electronics in the linear area.

By their very nature, such elements require the storage of energy in a volume. For high Q it is necessary that the volume be large. The incompatibility of large volume and integrated electronics is obvious from the terms themselves. Certain resonance phenomena, such as those in piezoelectric crystals, can be expected to have some applications for tuning functions, but inductors and capacitors will be with us for some time.

The integrated RF amplifier of the future might well consist of integrated stages of gain, giving high performance at minimum cost, interspersed with relatively large tuning elements.

Other linear functions will be changed considerably. The matching and tracking of similar components in integrated structures will allow the design of differential amplifiers of greatly improved performance. The use of thermal feedback effects to stabilize integrated structures to a small fraction of a degree will allow the construction of oscillators with crystal stability.

Even in the microwave area, structures included in the definition of integrated electronics will become increasingly important. The ability to make and assemble components small compared with the wavelengths involved will allow the use of lumped parameter design, at least at the lower frequencies. It is difficult to predict at the present time just how extensive the invasion of the microwave area by integrated electronics will be. The successful realization of such items as phased-array antennas, for example, using a multiplicity of integrated microwave power sources, could completely revolutionize radar.

APPENDIX C

 **G. E. Moore** is one of the new breed of electronic engineers, schooled in the physical sciences rather than in electronics. He earned a B.S. degree in chemistry from the University of California and a Ph.D. degree in physical chemistry from the California Institute of Technology. He was one of the founders of Fairchild Semiconductor and has been Director of the research and development laboratories since 1959.

D

Reprint of Robert H. Dennard's "Design of Ion-Implanted MOSFET's with Very Small Physical Dimensions"

NOTE: Reprinted from Robert H. Dennard, Fritz H. Gaensslen, Hwa-Nien Yu, V. Leo Rideout, Ernest Bassous, and Andre R. LeBlanc, 1974, Design of ion-implanted MOSFETS with very small physical dimensions, IEEE Journal of Solid State Circuits 9(5):256 with permission of IEEE and Robert H. Dennard © 1974 IEEE.

APPENDIX D

Design of Ion-Implanted MOSFET's with Very Small Physical Dimensions

ROBERT H. DENNARD, MEMBER, IEEE, FRITZ H. GAENSSLEN, HWA-NIEN YU, MEMBER, IEEE,
V. LEO RIDEOUT, MEMBER, IEEE, ERNEST BASSOUS, AND ANDRE R. LEBLANC, MEMBER, IEEE

Classic Paper

This paper considers the design, fabrication, and characterization of very small MOSFET switching devices suitable for digital integrated circuits using dimensions of the order of 1 μ. Scaling relationships are presented which show how a conventional MOSFET can be reduced in size. An improved small device structure is presented that uses ion implantation to provide shallow source and drain regions and a nonuniform substrate doping profile. One-dimensional models are used to predict the substrate doping profile and the corresponding threshold voltage versus source voltage characteristic. A two-dimensional current transport model is used to predict the relative degree of short-channel effects for different device parameter combinations. Polysilicon-gate MOSFET's with channel lengths as short as 0.5 μ were fabricated, and the device characteristics measured and compared with predicted values. The performance improvement expected from using these very small devices in highly miniaturized integrated circuits is projected.

I. LIST OF SYMBOLS

α	Inverse semilogarithmic slope of subthreshold characteristic.
D	Width of idealized step function profile for channel implant.
ΔW_f	Work function difference between gate and substrate.
$\epsilon_{Si}, \epsilon_{ox}$	Dielectric constants for silicon and silicon dioxide.
I_d	Drain current.
k	Boltzmann's constant.
κ	Unitless scaling constant.
L	MOSFET channel length.
μ_{eff}	Effective surface mobility.
n_i	Intrinsic carrier concentration.
N_a	Substrate acceptor concentration.
Ψ_s	Band bending in silicon at the onset of strong inversion for zero substrate voltage.
Ψ_b	Built-in junction potential.

This paper is reprinted from IEEE JOURNAL OF SOLID-STATE CIRCUITS, vol. SC-9, no. 5, pp. 256–268, October 1974.
Publisher Item Identifier S 0018-9219(99)02196-9.

q	Charge on the electron.
Q_{eff}	Effective oxide charge.
t_{ox}	Gate oxide thickness.
T	Absolute temperature.
V_d, V_s, V_g, V_{sub}	Drain, source, gate and substrate voltages.
V_{ds}	Drain voltage relative to source.
V_{s-sub}	Source voltage relative to sustrate.
V_t	Gate threshold voltage.
w_s, w_d	Source and drain depletion layer widths.
W	MOSFET channel width.

II. INTRODUCTION

New high resolution lithographic techniques for forming semiconductor integrated circuit patterns offer a decrease in linewidth of five to ten times over the optical contact masking approach which is commonly used in the semiconductor industry today. Of the new techniques, electron beam pattern writing has been widely used for experimental device fabrication [1]–[4] while X-ray lithography [5] and optical projection printing [6] have also exhibited high-resolution capability. Full realization of the benefits of these new high-resolution lithographic techniques requires the development of new device designs, technologies, and structures which can be optimized for very small dimensions.

This paper concerns the design, fabrication, and characterization of very small MOSFET switching devices suitable for digital integrated circuits using dimensions of the order of 1 μ. It is known that reducing the source-to-drain spacing (i.e., the channel length) of an FET leads to undesirable changes in the device characteristics. These changes become significant when the depletion regions surrounding the source and drain extend over a large portion of the region in the silicon substrate under the gate electrode. For switching applications, the most undesirable "short-channel" effect is a reduction in the gate threshold voltage at which the device turns on, which is aggravated

0018–9219/99$10.00 © 1999 IEEE

by high drain voltages. It has been shown that these short-channel effects can be avoided by scaling down the vertical dimensions (e.g., gate insulator thickness, junction depth, etc.) along with the horizontal dimensions, while also proportionately decreasing the applied voltages and increasing the substrate doping concentration [7], [8]. Applying this scaling approach to a properly designed conventional-size MOSFET shows that a 200-Å gate insulator is required if the channel length is to be reduced to 1 μ.

A major consideration of this paper is to show how the use of ion implantation leads to an improved design for very small scaled-down MOSFET's. First, the ability of ion implantation to accurately introduce a low concentration of doping atoms allows the substrate doping profile in the channel region under the gate to be increased in a controlled manner. When combined with a relatively lightly doped starting substrate, this channel implant reduces the sensitivity of the threshold voltage to changes in the source-to-substrate ("backgate") bias. This reduced "substrate sensitivity" can then be traded off for a thicker gate insulator of 350-Å thickness which tends to be easier to fabricate reproducibly and reliably. Second, ion implantation allows the formation of very shallow source and drain regions which are more favorable with respect to short-channel effects, while maintaining an acceptable sheet resistance. The combination of these features in an all-implanted design gives a switching device which can be fabricated with a thicker gate insulator if desired, which has well-controlled threshold characteristics, and which has significantly reduced interelectrode capacitances (e.g., drain-to-gate or drain-to-substrate capacitances).

This paper begins by describing the scaling principles which are applied to a conventional MOSFET to obtain a very small device structure capable of improved performance. Experimental verification of the scaling approach is then presented. Next, the fabrication process for an improved scaled-down device structure using ion implantation is described. Design considerations for this all-implanted structure are based on two analytical tools: a simple one-dimensional model that predicts the substrate sensitivity for long channel-length devices, and a two-dimensional current-transport model that predicts the device turn-on characteristics as a function of channel length. The predicted results from both analyses are compared with experimental data. Using the two-dimensional simulation, the sensitivity of the design to various parameters is shown. Then, detailed attention is given to an alternate design, intended for zero substrate bias, which offers some advantages with respect to threshold control. Finally, the paper concludes with a discussion of the performance improvements to be expected from integrated circuits that use these very small FET's.

III. DEVICE SCALING

The principles of device scaling [7], [8] show in a concise manner the general design trends to be followed in decreasing the size and increasing the performance of

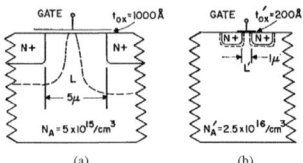

Fig. 1. Illustration of device scaling principles with $\kappa = 5$. (a) Conventional commercially available device structure. (b) Scaled-down device structure.

MOSFET switching devices. Fig. 1 compares a state-of-the-art n-channel MOSFET [9] with a scaled-down device designed following the device scaling principles to be described later. The larger structure shown in Fig. 1(a) is reasonably typical of commercially available devices fabricated by using conventional diffusion techniques. It uses a 1000-Å gate insulator thickness with a substrate doping and substrate bias chosen to give a gate threshold voltage V_t of approximately 2 V relative to the source potential. A substrate doping of 5×10^{15} cm^{-3} is low enough to give an acceptable value of substrate sensitivity. The substrate sensitivity is an important criterion in digital switching circuits employing source followers because the design becomes difficult if the threshold voltage increases by more than a factor of two over the full range of variation of the source voltage. For the device illustrated in Fig. 1(a), the design parameters limit the channel length L to about 5 μ. This restriction arises primarily from the penetration of the depletion region surrounding the drain into the area normally controlled by the gate electrode. For a maximum drain voltage of approximately 12–15 V this penetration will modify the surface potential and significantly lower the threshold voltage.

In order to design a new device suitable for smaller values of L, the device is scaled by a transformation in three variables: dimension, voltage, and doping. First, all linear dimensions are reduced by a unitless scaling factor κ, e.g., $t'_{\text{ox}} = t_{\text{ox}}/\kappa$, where the primed parameters refer to the new scaled-down device. This reduction includes vertical dimensions such as gate insulator thickness, junction depth, etc., as well as the horizontal dimensions of channel length and width. Second, the voltages applied to the device are reduced by the same factor (e.g., $V'_{ds} = V_{ds}/\kappa$). Third, the substrate doping concentration is increased, again using the same scaling factor (i.e., $N'_a = \kappa N_a$). The design shown in Fig. 1(b) was obtained using $\kappa = 5$ which corresponds to the desired reduction in channel length to 1 μ.

The scaling relationships were developed by observing that the depletion layer widths in the scaled-down device are reduced in proportion to the device dimensions due to the reduced potentials and the increased doping. For example

$$w'_s = \{[2\epsilon_{\text{Si}}(\psi'_b + V_{s-\text{sub}}/\kappa)]/q\kappa N_a\}^{1/2} \simeq w_s/\kappa. \quad (1)$$

The threshold voltage at turn-on [9] is also decreased in direct proportion to the reduced device voltages so that

APPENDIX D

the device will function properly in a circuit with reduced voltage levels. This is shown by the threshold voltage equation for the scaled-down device

$$V_t' = (t_{ox}/\kappa\epsilon_{ox})$$
$$\cdot \left\{ -Q_{eff} + [2\epsilon_{Si}q\kappa N_a(\psi_s' + V_{s-sub}/\kappa)]^{1/2} \right\}$$
$$+ (\Delta W_f + \psi_s') \simeq V_t/\kappa. \quad (2)$$

In (2) the reduction in V_t is primarily due to the decreased insulator thickness, t_{ox}/κ, while the changes in the voltage and doping terms tend to cancel out. In most cases of interest (i.e., polysilicon gates of doping type opposite to that of the substrate or aluminum gates on p-type substrates) the work function difference ΔW_f is of opposite sign, and approximately cancels out ψ_s'. ψ_s' is the band bending in the silicon (i.e., the surface potential) at the onset of strong inversion for zero substrate bias. It would appear that the ψ' terms appearing in (1) and (2) prevent exact scaling since they remain approximately constant, actually increasing slightly due to the increased doping since $\psi_b' \simeq \psi_s' = (2kT/q)\ln(N_a'/n_i)$. However, the fixed substrate bias supply normally used with n-channel devices can be adjusted so that $(\psi_s' + V_{sub}') = (\psi_s + V_{sub})/\kappa$. Thus, by scaling down the applied substrate bias more than the other applied voltages, the potential drop across the source or drain junctions, or across the depletion region under the gate, can be reduced by κ.

All of the equations that describe the MOSFET device characteristics may be scaled as demonstrated above. For example, the MOSFET current equation [9] given by

$$I_d' = \frac{\mu_{eff}\epsilon_{ox}}{t_{ox}/\kappa}\left(\frac{W/\kappa}{L/\kappa}\right)\left(\frac{V_g - V_t - V_d/2}{\kappa}\right)(V_d/\kappa) = I_d/\kappa \quad (3)$$

is seen to be reduced by a factor of κ, for any given set of applied voltages, assuming no change in mobility. Actually, the mobility is reduced slightly due to increased impurity scattering in the heavier doped substrate.

It is possible to generalize the scaling approach to include electric field patterns and current density. The electric field distribution is maintained in the scaled-down device except for a change in scale for the spatial coordinates. Furthermore, the electric field strength at any corresponding point is unchanged because $V/x = V'/x'$. Thus, the carrier velocity at any point is also unchanged due to scaling and, hence, any saturation velocity effects will be similar in both devices, neglecting microscopic differences due to the fixed crystal lattice dimensions. From (3), since the device current is reduced by κ, the channel current per unit of channel width W is unchanged by scaling. This is consistent with the same sheet density of carriers (i.e., electrons per unit gate area) moving at the same velocity. In the vicinity of the drain the carriers will move away from the surface to a lesser extent in the new device, due to the shallower diffusions. Thus, the density of mobile carriers per unit volume will be higher in the space-charge region around the drain complementing the higher density of immobile charge due to the heavier doped substrate. Other scaling

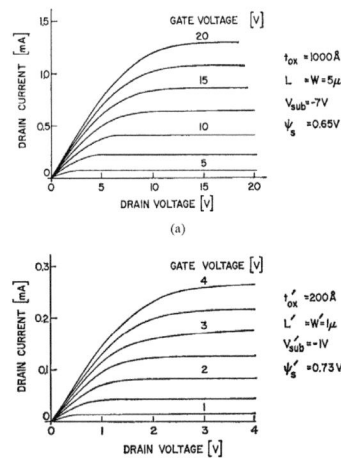

Fig. 2. Experimental drain voltage characteristics for (a) conventional, and (b) scaled-down structures shown in Fig. 1 normalized to $W/L = 1$.

relationships for power density, delay time, etc., are given in Table 1 and will be discussed in a subsequent section on circuit performance.

In order to verify the scaling relationships, two sets of experimental devices were fabricated with gate insulator of 1000 and 200 Å (i.e., $\kappa = 5$). The measured drain voltage characteristics of these devices, normalized to $W/L = 1$ are shown in Fig. 2. The two sets of characteristics are quite similar when plotted with voltage and current scale of the smaller device reduced by a factor of five, which confirms the scaling predictions. In Fig. 2, the exact match on the current scale is thought to be fortuitous since there is some experimental uncertainty in the magnitude of the channel length used to normalize the characteristics (see Appendix). More accurate data from devices with larger width and length dimensions on the same chip shows an approximate reduction of ten percent in mobility for devices with the heavier doped substrate. That the threshold voltage also scales correctly by a factor of five is verified in Fig. 3, which shows the experimental $\sqrt{I_d}$ versus V_g turn-on characteristics for the original and the scaled-down devices. For the cases shown, the drain voltage is large enough to cause pinchoff and the characteristics exhibit the expected linear relationship. When projected to intercept the gate voltage axis this linear relationship defines a threshold voltage useful for most logic circuit design purposes.

One area in which the device characteristics fail to scale is in the subthreshold or weak inversion region of the turn-on characteristic. Below threshold, I_d is exponentially dependent on V_g with an inverse semilogarithmic slope, α,

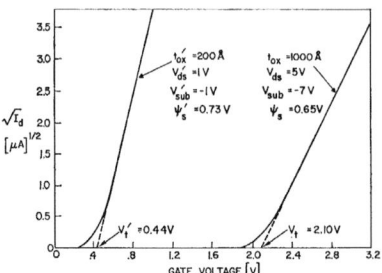

Fig. 3. Experimental turn-on characteristics for conventional and scaled-down devices shown in Fig. 1 normalized to $W/L = 1$.

[10], [11] which for the scaled-down device is given by

$$\alpha'\left(\frac{\text{volts}}{\text{decade}}\right) = \frac{dV_g'}{d\log_{10} I_d'}$$
$$= (kT/q \log_{10} e)\left(1 + \frac{\epsilon_{\text{Si}} t_{\text{ox}}/\kappa}{\epsilon_{\text{ox}} w_d/\kappa}\right) \quad (4)$$

which is the same as for the original larger device. The parameter α is important to dynamic memory circuits because it determines the gate voltage excursion required to go from the low current "off" state to the high current "on" state [11]. In an attempt to also extend the linear scaling relationships to α one could reduce the operating temperature in (4) (i.e., $T' = T/\kappa$), but this would cause a significant increase in the effective surface mobility [12] and thereby invalidate the current scaling relationship of (3). In order to design devices for operation at room temperature and above, one must accept the fact that the subthreshold behavior does not scale as desired. This nonscaling property of the subthreshold characteristic is of particular concern to miniature dynamic memory circuits which require low source-to-drain leakage currents.

IV. ION-IMPLANTED DEVICE DESIGN

The scaling considerations just presented lead to the device structure with a 1-μ channel length shown in Fig. 4(a). In contrast, the corresponding improved design utilizing the capability afforded by ion implantation is shown in Fig. 4(b). The ion-implanted device uses an initial substrate doping that is lower by about a factor of four, and an implanted boron surface layer having a concentration somewhat greater than the concentration used throughout the unimplanted structure of Fig. 4(a). The concentration and the depth of the implanted surface layer are chosen so that this heavier doped region will be completely within the surface depletion layer when the device is turned on with the source grounded. Thus, when the source is biased above ground potential, the depletion layer will extend deeper into the lighter doped substrate, and the additional exposed "bulk" charge will be reasonably small and will cause only

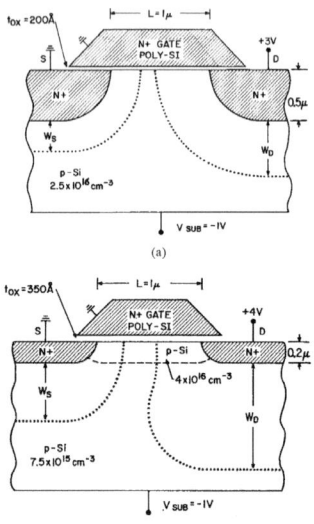

Fig. 4. Detailed cross sections for (a) scaled-down device structure, and (b) corresponding ion-implanted device structure.

a modest increase in the gate-to-source voltage required to turn on the device. With this improvement in substrate sensitivity the gate insulator thickness can be increased to as much as 350 Å and still maintain a reasonable gate threshold voltage as will be shown later.

Another aspect of the design philosophy is to use shallow implanted n$^+$ regions of depth comparable to the implanted p-type surface layer. The depletion regions under the gate electrode at the edges of the source and drain are then inhibited by the heavier doped surface layer, roughly pictured in Fig. 4(b), for the case of a turned-off device. The depletion regions under the source and drain extend much further into the lighter doped substrate. With deeper junctions these depletion regions would tend to merge in the lighter doped material which would cause a loss of threshold control or, in the extreme, punchthrough at high drain voltages. However, the shallower junctions give a more favorable electric field pattern which avoids these effects when the substrate doping concentration is properly chosen (i.e., when it is not too light).

The device capacitances are reduced with the ion-implanted structure due to the increased depletion layer width separating the source and drain from the substrate [cf. Figs. 4(a) and 4(b)], and due to the natural self-alignment afforded by the ion implantation process which reduces the overlap of the polysilicon gate over the source and drain regions. The thicker gate insulator also gives reduced gate capacitance, but the performance benefit in this respect is

APPENDIX D

offset by the decreased gate field. To compensate for the thicker gate oxide and the expected threshold increase, a design objective for maximum drain voltage was set at 4 V for the ion-implanted design in Fig. 4(b), compared to 3 V for the scaled-down device of Fig. 4(a).

V. FABRICATION OF ION-IMPLANTED MOSFET'S

The fabrication process for the ion-implanted MOSFET's used in this study will now be described. A four-mask process was used to fabricate polysilicon-gate, n-channel MOSFET's on a test chip which contains devices with channel lengths ranging from 0.5 to 10 μ. Though the eventual aim is to use electron-beam pattern exposure, it was more convenient to use contact masking with high quality master masks for process development. For this purpose high resolution is required only for the gate pattern which uses lines as small as 1.5 μ which are reduced in the subsequent processing. The starting substrate resistivity was 2 $\Omega \cdot$ cm (i.e., about 7.5×10^{15} cm^{-3}). The method of fabrication for the thick oxide isolation between adjacent FET's is not described as it is not essential to the work presented here, and because several suitable techniques are available. Following dry thermal growth of the gate oxide, low energy (40 keV), low dose (6.7×10^{11} atoms/cm^2) B^{11} ions were implanted into the wafers, raising the boron doping near the silicon surface. All implantations were performed after gate oxide growth in order to restrict diffusion of the implanted regions.

After the channel implantation, a 3500-Å thick polysilicon layer was deposited, doped n$^+$, and the gate regions delineated. Next, n$^+$ source and drain regions 2000-Å deep were formed by a high energy (100 keV), high dose (4×10^{15} atoms/cm^2) As^{75} implantation through the same 350-Å oxide layer. During this step, however, the polysilicon gate masks the channel region from the implant, absorbing all of the As^{75} dose incident there. The etching process used to delineate the gates results in a sloping sidewall which allows a slight penetration of As^{75} ions underneath the edges of the gates. The gate-to-drain (or source) overlap is estimated to be on the order of 0.2 μ. The high temperature processing steps that follow the implantations include 20 min at 900°C, and 11 min at 1000°C, which is more than adequate to anneal out the implantation damage without greatly spreading out the implanted doses. Typical sheet resistances were 50 $\Omega/\square$ for the source and drain regions, and 40 $\Omega/\square$ for the polysilicon areas. Following the As^{75} implant, a final insulating oxide layer 2000-Å thick was deposited using low-temperature chemical-vapor deposition. Then, the contact holes to the n$^+$ and polysilicon regions were defined, and the metalization was applied and delineated. Electrical contact directly to the shallow implanted source and drain regions was accomplished by a suitably chosen metallurgy to avoid junction penetration due to alloying during the final annealing step. After metalization an annealing step of 400°C for 20 min in forming gas was performed to decrease the fast-state density.

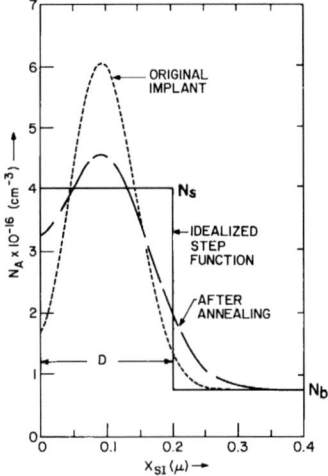

Fig. 5. Predicted substrate doping profile for basic ion-implanted device design for 40 keV B^{11} ions implanted through the 350-Å gate insulator.

VI. ONE-DIMENSIONAL (LONG CHANNEL) ANALYSIS

The substrate doping profile for the 40 keV, 6.7×10^{11} atoms/cm^2 channel implant incident on the 350-Å gate oxide, is shown in Fig. 5. Since the oxide absorbs 3 percent of the incident dose, the active dose in the silicon is 6.5×10^{11} atoms/cm^2. The concentration at the time of the implantation is given by the lightly dashed Gaussian function added to the background doping level, N_b. For 40 keV B^{11} ions, the projected range and standard deviation were taken as 1300 Å and 500 Å, respectively [13]. After the heat treatments of the subsequent processing, the boron is redistributed as shown by the heavier dashed line. These predicted profiles were obtained using a computer program developed by F. F. Morehead of our laboratories. The program assumes that boron atoms diffusing in the silicon reflect from the silicon-oxide interface and thereby raise the surface concentration. For modeling purposes it is convenient to use a simple, idealized, step-function representation of the doping profile, as shown by the solid line in Fig. 5. The step profile approximates the final predicted profile rather well and offers the advantage that it can be described by a few simple parameters. The three profiles shown in Fig. 5 all have the same active dose.

Using the step profile, a model for determining threshold voltage has been developed from piecewise solutions of Poisson's equation with appropriate boundary conditions [11]. The one-dimensional model considers only the vertical dimension and cannot account for horizontal short-channel effects. Results of the model are shown in Fig. 6 which plots the threshold voltage versus source-to-substrate bias for the ion-implanted step profile shown in Fig. 5. For comparison, Fig. 6 also shows the substrate sensitivity characteristics for the nonimplanted device with a 200-Å gate insulator and a constant background doping, and for a hypothetical device having a 350-Å gate insulator like the implanted structure and a constant background doping like the nonimplanted structure. The nonimplanted 200-Å case exhibits a low substrate sensitivity, but the magnitude of the threshold voltage is also low. On the other hand, the nonimplanted 350-Å case shows a higher threshold, but with an undesirably high substrate sensitivity. The ion-implanted case offers both a sufficiently high threshold voltage and a reasonably low substrate sensitivity, particularly for $V_{s-\text{sub}} \geq 1$ V. For $V_{s-\text{sub}} < 1$ V, a steep slope occurs because the surface inversion layer in the channel is obtained while the depletion region in the silicon under the gate does not exceed D, the step width of the heavier doped implanted region. For $V_{s-\text{sub}} \geq 1$ V, at inversion the depletion region now extends into the lighter doped substrate and the threshold voltage then increases relatively slowly with $V_{s-\text{sub}}$ [11]. Thus, with a fixed substrate bias of -1 V, the substrate sensitivity over the operating range of the source voltage (e.g., ground potential to 4 V) is reasonably low and very similar to the slope of the nonimplanted 200-Å design. However, the threshold voltage is significantly higher for the implanted design which allows adequate design margin so that, under worst case conditions (e.g., short-channel effects which reduce the threshold considerably), the threshold will still be high enough so that the device can be turned off to a negligible conduction level as required for dynamic memory applications.

Experimental results are also given in Fig. 6 from measurements made on relatively long devices (i.e., $L = 10\ \mu$) which have no short-channel effects. These data agree reasonably well with the calculated curve. A 35 keV, 6×10^{11} atoms/cm^2 implant was used to achieve this result, rather than the slightly higher design value of 40 keV and 6.7×10^{11} atoms/cm^2.

VII. Two-Dimensional (Short Channel) Analysis

For devices with sufficiently short-channel lengths, the one-dimensional model is inadequate to account for the threshold voltage lowering due to penetration of the drain field into the channel region normally controlled by the gate. While some models have been developed which account for this behavior [14], the problem is complicated for the ion-implanted structure by the nonuniform doping profile which leads to an electric field pattern that is difficult to approximate. For the ion-implanted case, the two-dimensional numerical current transport model of Kennedy and Mock [15], [16] was utilized. The computer program was modified by W. Chang and P. Hwang [17] to handle the abrupt substrate doping profiles considered for these devices.

The numerical current transport model was used to calculate the turn-on behavior of the ion-implanted device

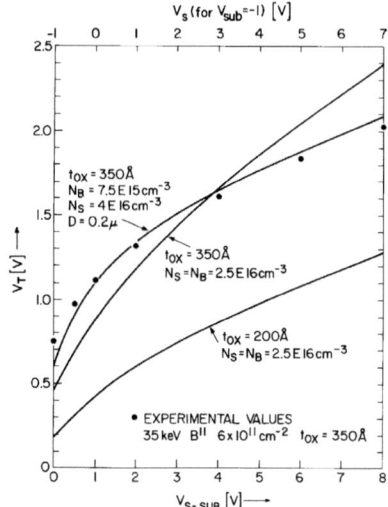

Fig. 6. Calculated and experimental substrate sensitivity characteristics for non-implanted devices with 200- and 350-Å gate insulators, and for corresponding ion-implanted device with 350-Å gate insulator.

by a point-by-point computation of the device current for increasing values of gate voltage. Calculated results are shown in Fig. 7 for two values of channel length in the range of 1 μ, as well as for a relatively long-channel device with $L = 10\ \mu$. All cases were normalized to a width-to-length ratio of unity, and a drain voltage of 4 V was used in all cases. As the channel length is reduced to the order of 1 μ, the turn-on characteristic shifts to a lower gate voltage due to a lowering of the threshold voltage. The threshold voltage occurs at about 10^{-7} A where the turn-on characteristics make a transition from the exponential subthreshold behavior (a linear response on this semilogarithmic plot) to the $I_d \propto V_g^2$ square-law behavior. This current level can also be identified from Fig. 3 as the actual current at the projected threshold voltage, V_t. When the computed characteristics were plotted in the manner of Fig. 3 they gave 4×10^{-8} A at threshold for all device lengths. The band bending, ψ_s, at this threshold condition is approximately 0.75 V. Some of the other device designs considered with heavier substrate concentrations gave a higher current at threshold, so, for simplicity, the value of 10^{-7} A was used in all cases with a resultant small error in V_t.

MOSFET's with various channel lengths were measured to test the predictions of the two-dimensional model. The technique for experimentally determining the channel length for very short devices is described in the Appendix.

APPENDIX D

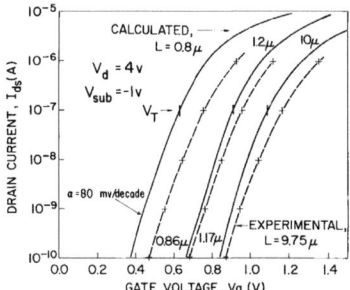

Fig. 7. Calculated and experimental subthreshold turn-on characteristic for basic ion-implanted design for various channel lengths with $V_{\text{sub}} = -1$ V, $V_d = 4$ V.

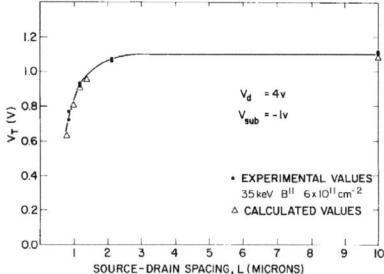

Fig. 8. Experimental and calculated dependence of threshold voltage on channel length for basic ion-implanted design with $V_{\text{sub}} = -1$ V, $V_d = 4$ V.

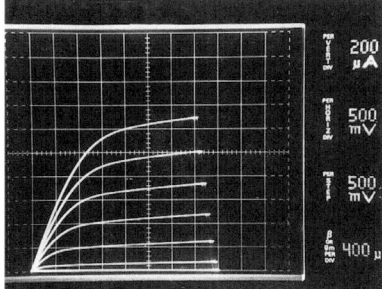

Fig. 9. Experimental drain voltage characteristics for basic ion-implanted design with $V_{\text{sub}} = -1$ V, $L = 1.1\ \mu$, and $W = 12.2\ \mu$. Curve tracer parameters; load resistance 30 Ω, drain voltage 4 V, gate voltage 0–4 in 8 steps each 0.5 V apart.

The experimental results are plotted in Fig. 7 and show good agreement with the calculated curves, especially considering the somewhat different values of L. Another form of presentation of this data is shown in Fig. 8 where the threshold voltage is plotted as a function of channel length. The threshold voltage is essentially constant for $L > 2\ \mu$, and falls by a reasonably small amount as L is decreased from 2 to 1 μ, and then decreases more rapidly with further reductions in L. For circuit applications the nominal value of L could be set somewhat greater than 1 μ so that, over an expected range of deviation of L, the threshold voltage is reasonably well controlled. For example, $L = 1.3 \pm 0.3\ \mu$ would give $V_t = 1.0 \pm 0.1$ V from chip to chip due to this short-channel effect alone. This would be tolerable for many circuit applications because of the tracking of different devices on a given chip, if indeed this degree of control of L can be achieved. The experimental drain characteristics for an ion-implanted MOSFET with a 1.1-μ channel length are shown in Fig. 9 for the grounded source condition. The general shape of the characteristics is the same as those observed for much larger devices. No extraneous short-channel effects were observed for drain voltages as large as 4 V. The experimental data in Figs. 6–9 were taken from devices using a B^{11} channel implantation energy and dose of 35 keV and 6.0×10^{11} atoms/cm^2, respectively.

The two-dimensional simulations were also used to test the sensitivity of the design to various parameters. The results are given in Fig. 10 which tabulates values of threshold voltage as a function of channel length for the indicated voltages. Fig. 10(a) is an idealized representation for the basic design that has been discussed thus far. The first perturbation to the basic design was an increase in junction depth to 0.4 μ. This was found to give an appreciable reduction in threshold voltage for the shorter devices in Fig. 10(b). Viewed another way, the minimum device length would have to be increased by 20 percent (from 1.0 to 1.2 μ) to obtain a threshold comparable to the basic design. This puts the value of the shallower junctions in perspective. Another perturbation from the basic design which was considered was the use of a substrate doping lighter by a factor of 2, with a slightly higher concentration in the surface layer to give the same threshold for a long-channel device [Fig. 10(c)]. The results for smaller devices proved to be similar to the case of deeper junctions. The next possible departure from the basic design is the use of a shallower boron implantation in the channel region, only half as deep, with a heavier concentration to give the same long-channel threshold [Fig. 10(d)]. With the shallower profile, and considering that the boron dose implanted in the silicon is about 20 percent less in this case, it was expected that more short-channel effects would occur. However, the calculated values show almost identical thresholds compared to the basic design. With the shallower implantation it is possible to use zero substrate bias and still have good substrate sensitivity since the heavier doped region is completely depleted at turn-on with a grounded source. The last design perturbation considers such a case, again with a heavier concentration to give the same long-

182 THE FUTURE OF COMPUTING PERFORMANCE

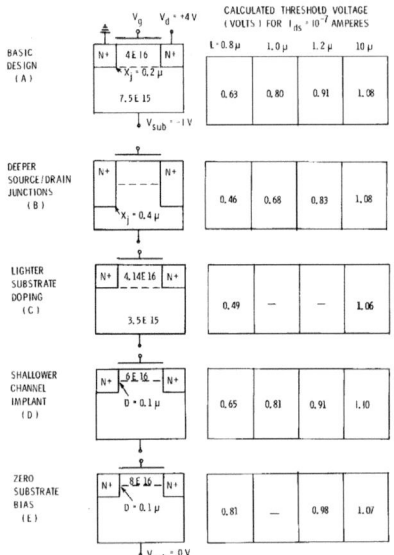

Fig. 10. Threshold voltage calculated using two-dimensional current transport model for various parameter conditions. A flatband voltage of -1.1 V is assumed.

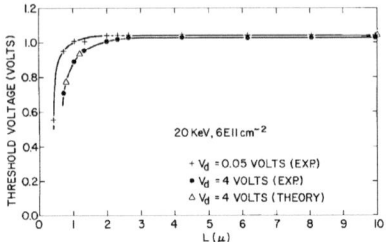

Fig. 11. Experimental and calculated dependence of threshold voltage on channel length for ion-implanted zero substrate bias design.

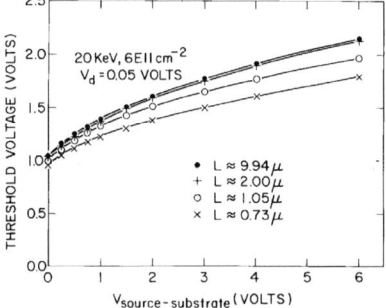

Fig. 12. Substrate sensitivity characteristics for ion-implanted zero substrate bias design with channel length as parameter.

channel threshold [Fig. 10(e)]. The calculations for this case show appreciably less short-channel effect. In fact, the threshold for this case for a device with $L = 0.8~\mu$ is about the same as for an $L = 1.0~\mu$ device of the basic design. This important improvement is apparently due to the reduced depletion layer widths around the source and drain with the lower voltage drop across those junctions. Also, with these bias and doping conditions, the depletion layer depth in the silicon under the gate is much less at threshold, particularly near the source where only the band bending, ψ_s, appears across this depletion region, which may help prevent the penetration of field lines from the drain into this region where the device turn-on is controlled.

VIII. CHARACTERISTICS OF THE ZERO SUBSTRATE BIAS DESIGN

Since the last design shown in Fig. 10(e) appears to be better behaved in terms of short-channel effects, it is worthwhile to review its properties more fully. Experimental devices corresponding to this design were built and tested with various channel lengths. In this case a 20 keV, 6.0×10^{11} atoms/cm^2 B^{11} implant was used to obtain a shallower implanted layer of approximately 1000-Å depth [11]. Data on threshold voltage for these devices with 4 V applied to the drain is presented in Fig. 11 and corresponds very well to the calculated values. Data for a small drain voltage is also given in this figure, showing much less variation of threshold with channel length, as expected. The dependence of threshold voltage on source-to-substrate bias is shown in Fig. 12 for different values of L. The drain-to-source voltage was held at a constant low value for this measurement. The results show that the substrate sensitivity is indeed about the same for this design with zero substrate bias as for the original design with $V_{\text{sub}} = -1$ V. Note that the smaller devices show a somewhat flatter substrate sensitivity characteristic with relatively lower thresholds at high values of source (and drain) voltage.

The turn-on characteristics for the zero substrate bias design, both experimental and calculated, are shown in Fig. 13 for different values of L. The relatively small shift in threshold for the short-channel devices is evident; however, the turn-on rate is considerably slower for this case than for the $V_{\text{sub}} = -1$ V case shown in Fig. 7. This is due to the fact that the depletion region in the silicon under the gate is very shallow for this zero substrate bias case so that a large portion of a given gate voltage change is dropped across the gate insulator capacitance rather than across the silicon depletion layer capacitance. This is discussed in some detail

APPENDIX D

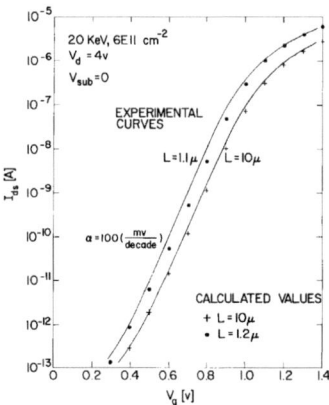

Fig. 13. Calculated and experimental subthreshold turn-on characteristics for ion-implanted zero substrate bias design.

for these devices in another paper [11]. The consequence for dynamic memory applications is that, even though the zero substrate bias design offers improved threshold control for strong inversion, this advantage is offset by the flatter subthreshold turn-on characteristic. For such applications the noise margin with the turn-on characteristic of Fig. 13 is barely suitable if the device is turned off by bringing its gate to ground. Furthermore, elevated temperature aggravates the situation [18]. Thus, for dynamic memory, the basic design with $V_{sub} = -1$ V presented earlier is preferred.

IX. CIRCUIT PERFORMANCE WITH SCALED-DOWN DEVICES

The performance improvement expected from using very small MOSFET's in integrated circuits of comparably small dimensions is discussed in this section. First, the performance changes due to size reduction alone are obtained from the scaling considerations given earlier. The influence on the circuit performance due to the structural changes of the ion-implanted design is then discussed.

Table 1 lists the changes in integrated circuit performance which follow from scaling the circuit dimensions, voltages, and substrate doping in the same manner as the device changes described with respect to Fig. 1. These changes are indicated in terms of the dimensionless scaling factor κ. Justifying these results here in great detail would be tedious, so only a simplified treatment is given. It is argued that all nodal voltages are reduced in the miniaturized circuits in proportion to the reduced supply voltages. This follows because the quiescent voltage levels in digital MOSFET circuits are either the supply levels or some intermediate level given by a voltage divider consisting of two or more devices, and because the resistance V/I of each device is unchanged by scaling. An assumption is made that

Table 1
Scaling Results for Circuit Performance

Device or Circuit Parameter	Scaling Factor
Device dimension t_{ox}, L, W	$1/\kappa$
Doping concentration N_a	κ
Voltage V	$1/\kappa$
Current I	$1/\kappa$
Capacitance $\epsilon A/t$	$1/\kappa$
Delay time/circuit VC/I	$1/\kappa$
Power dissipation/circuit VI	$1/\kappa^2$
Power density VI/A	1

parasitic resistance elements are either negligible or unchanged by scaling, which will be examined subsequently. The circuits operate properly at lower voltages because the device threshold voltage V_t scales as shown in (2), and furthermore because the tolerance spreads on V_t should be proportionately reduced as well if each parameter in (2) is controlled to the same percentage accuracy. Noise margins are reduced, but at the same time internally generated noise coupling voltages are reduced by the lower signal voltage swings.

Due to the reduction in dimensions, all circuit elements (i.e., interconnection lines as well as devices) will have their capacitances reduced by a factor of κ. This occurs because of the reduction by κ^2 in the area of these components, which is partially cancelled by the decrease in the electrode spacing by κ due to thinner insulating films and reduced depletion layer widths. These reduced capacitances are driven by the unchanged device resistances V/I giving decreased transition times with a resultant reduction in the delay time of each circuit by a factor of κ. The power dissipation of each circuit is reduced by κ^2 due to the reduced voltage and current levels, so the power-delay product is improved by κ^3. Since the area of a given device or circuit is also reduced by κ^2, the power density remains constant. Thus, even if many more circuits are placed on a given integrated circuit chip, the cooling problem is essentially unchanged.

As indicated in Table 2, a number of problems arise from the fact that the cross-sectional area of conductors is decreased by κ^2 while the length is decreased only by κ. It is assumed here that the thicknesses of the conductors are necessarily reduced along with the widths because of the more stringent resolution requirements (e.g., on etching, etc.). The conductivity is considered to remain constant which is reasonable for metal films down to very small dimensions (until the mean free path becomes comparable to the thickness), and is also reasonable for degenerately doped semiconducting lines where solid solubility and impurity scattering considerations limit any increase in conductivity. Under these assumptions the resistance of a given line increases directly with the scaling factor κ. The IR drop in such a line is therefore constant (with the decreased current levels), but is κ times greater in comparison to the lower operating voltages. The response time of an unterminated transmission line is characteristically limited

Table 2
Scaling Results for Interconnection Lines

Parameter	Scaling Factor
Line resistance, $R_L = \rho L/Wt$	κ
Normalized voltage drop IR_L/V	κ
Line response time $R_L C$	1
Line current density I/A	κ

by its time constant $R_L C$, which is unchanged by scaling; however, this makes it difficult to take advantage of the higher switching speeds inherent in the scaled-down devices when signal propagation over long lines is involved. Also, the current density in a scaled-down conductor is increased by κ, which causes a reliability concern. In conventional MOSFET circuits, these conductivity problems are relatively minor, but they become significant for linewidths of micron dimensions. The problems may be circumvented in high performance circuits by widening the power buses and by avoiding the use of n⁺ doped lines for signal propagation.

Use of the ion-implanted devices considered in this paper will give similar performance improvement to that of the scaled-down device with $\kappa = 5$ given in Table 1. For the implanted devices with the higher operating voltages (4 V instead of 3 V) and higher threshold voltages (0.9 V instead of 0.4 V), the current level will be reduced in proportion to $(V_g - V_t)^2/t_{ox}$ to about 80 percent of the current in the scaled-down device. The power dissipation per circuit is thus about the same in both cases. All device capacitances are about a factor of two less in the implanted devices, and n⁺ interconnection lines will show the same improvement due to the lighter substrate doping and decreased junction depth. Some capacitance elements such as metal interconnection lines would be essentially unchanged so that the overall capacitance improvement in a typical circuit would be somewhat less than a factor of two. The delay time per circuit which is proportional to VC/I thus appears to be about the same for the implanted and for the directly scaled-down micron devices shown in Fig. 4.

X. Summary

This paper has considered the design, fabrication, and characterization of very small MOSFET switching devices. These considerations are applicable to highly miniaturized integrated circuits fabricated by high-resolution lithographic techniques such as electron-beam pattern writing. A consistent set of scaling relationships were presented that show how a conventional device can be reduced in size; however, this direct scaling approach leads to some challenging technological requirements such as very thin gate insulators. It was then shown how an all ion-implanted structure can be used to overcome these difficulties without sacrificing device area or performance. A two-dimensional current transport model modified for use with ion-implanted structures proved particularly valuable in predicting the relative degree of short-channel effects arising from different device parameter combinations. The general objective of the study was to design an n-channel polysilicon-gate MOSFET with a 1-μ channel length for high-density source-follower circuits such as those used in dynamic memories. The most satisfactory combination of subthreshold turn-on range, threshold control, and substrate sensitivity was achieved by an experimental MOSFET that used a 35 keV, 6.0×10^{11} atoms/cm² B¹¹ channel implant, a 100 keV, 4×10^{15} atoms/cm² As⁷⁵ source/drain implant, a 350-Å gate insulator, and an applied substrate bias of -1 V. Also presented was an ion-implanted design intended for zero substrate bias that is more attractive from the point of view of threshold control but suffers from an increased subthreshold turn-on range. Finally the sizable performance improvement expected from using very small MOSFET's in integrated circuits of comparably small dimensions was projected.

Appendix
Experimental Determination of Channel Length

A technique for determining the effective electrical channel length L for very small MOSFET's from experimental data is described here. The technique is based on the observation that

$$WR_{\text{chan}} = L\rho_{\text{chan}} \quad (A1)$$

where R_{chan} is the channel resistance, and ρ_{chan} the sheet resistance of the channel. For a fixed value of $V_g - V_t > 0$, and with the device turned on in the below-pinchoff region, the channel sheet resistance is relatively independent of L. Then, a plot of WR_{chan} versus L_{mask} will intercept the L_{mask} axis at ΔL because $\Delta L = L_{\text{mask}} - L$, where ΔL is the processing reduction in the mask dimension due to exposure and etching. An example of this technique is illustrated in Fig. 14.

The experimental values of W and R_{chan} used in Fig. 14 were obtained as follows. First, the sheet resistance of the ion-implanted n⁺ region was determined using a relatively large four-point probe structure. Knowing the n⁺ sheet resistance allows us to compute the source and drain resistance R_s and R_d, and to deduce W from the resistance of a long, slender, n⁺ line. The channel resistance can be calculated from

$$R_{\text{chan}} = V_{\text{chan}}/I_d = (V_d - I_d(R_s + R_d + 2R_c + R_{\text{load}}))/I_d \quad (A2)$$

where R_c is the contact resistance of the source or drain, and R_{load} is the load resistance of the measurement circuit. I_d was determined at $V_g = V_t + 0.5$ V with a small applied drain voltage of 50 or 100 mV. The procedure is more simple and accurate if one uses a set of MOSFET's having different values of L_{mask} but all with the same value of

APPENDIX D

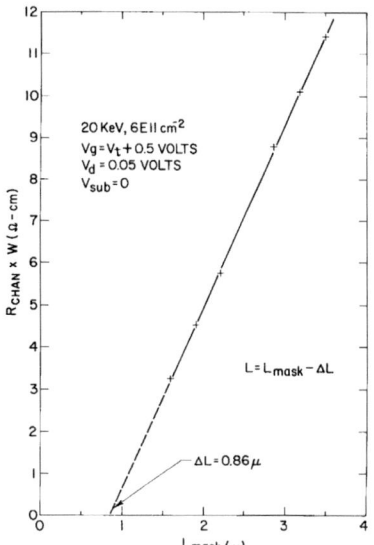

Fig. 14. Illustration of experimental technique used to determine channel length, L.

W_{mask}. Then one needs only to plot R_{chan} versus L_{mask} in order to determine ΔL.

ACKNOWLEDGMENT

We wish to acknowledge the valuable contributions of B. L. Crowder and F. F. Morehead who provided the ion implantations and related design information. Also important were the contributions of P. Hwang and W. Chang to two-dimensional device computations. J. J. Walker and V. DiLonardo assisted with the mask preparation and testing activities. The devices were fabricated by the staff of the silicon technology facility at the T. J. Watson Research Center.

REFERENCES

[1] F. Fang, M. Hatzakis, and C. H. Ting, "Electron-beam fabrication of ion implanted high-performance FET circuits," *J. Vac. Sci. Technol.*, vol. 10, p. 1082, 1973.
[2] J. M. Pankrantz, H. T. Yuan, and L. T. Creagh, "A high-gain, low-noise transistor fabricated with electron beam lithography," in *Tech. Dig. Int. Electron Devices Meeting*, Dec. 1973, pp. 44–46.
[3] H. N. Yu, R. H. Dennard, T. H. P. Chang, and M. Hatzakis, "An experimental high-density memory array fabricated with electron beam," in *ISSCC Dig. Tech. Papers*, Feb. 1973, pp. 98–99.
[4] R. C. Henderson, R. F. W. Pease, A. M. Voshchenkow, P. Mallery, and R. L. Wadsack, "A high speed p-channel random access 1024-bit memory made with electron lithography," in *Tech. Dig. Int. Electron Devices Meeting*, Dec. 1973, pp. 138–140.
[5] D. L. Spears and H. I. Smith, "X-Ray lithography—a new high resolution replication process," *Solid State Technol.*, vol. 15, p. 21, 1972.
[6] S. Middlehoek, "Projection masking, thin photoresist layers and interference effects," *IBM J. Res. Develop.*, vol. 14, p. 117, 1970.
[7] R.*H. Dennard, F. H. Gaensslen, L. Kuhn, and H. N. Yu, "Design of micron MOS switching devices," presented at the IEEE Int. Electron Devices Meeting, Washington, D.C., Dec. 1972.
[8] A. N. Broers and R. H. Dennard, "Impact of electron beam technology on silicon device fabrication," *Semicond. Silicon* (Electrochem. Soc. Publication), H. R. Huff and R. R. Burgess, eds., pp. 830–841, 1973.
[9] D. L. Critchlow, R. H. Dennard, and S. E. Schuster, "Design characteristics of n-channel insulated-gate field-effect transistors," *IBM J. Res. Develop.*, vol. 17, p. 430, 1973.
[10] R. M. Swanson and J. D. Meindl, "Ion-implanted complementary MOS transistors in low-voltage circuits," *IEEE J. Solid-State Circuits*, vol. SC-7, pp. 146–153, April 1972.
[11] V. L. Rideout, F. H. Gaensslen, and A. LeBlanc, "Device design considerations for ion implanted n-channel MOSFET's," *IBM J. Res. Develop.*, to be published.
[12] F. F. Fang and A. B. Fowler, "Transport properties of electrons in inverted Si surfaces," *Phys. Rev.*, vol. 169, p. 619, 1968.
[13] W. S. Johnson, IBM System Products Division, E. Fishkill, N.Y., private communication.
[14] H. S. Lee, "An analysis of the threshold voltage for short channel IGFET's," *Solid-State Electron.*, vol. 16, p. 1407, 1973.
[15] D. P. Kennedy and P. C. Murley, "Steady state mathematical theory for the insulated gate field effect transistor," *IBM J. Res. Develop.*, vol. 17, p. 1, 1973.
[16] M. S. Mock, "A two-dimensional mathematical model of the insulated-gate field-effect transistor," *Solid-State Electron.*, vol. 16, p. 601, 1973.
[17] W. Chang and P. Hwang, IBM System Products Division, Essex Junction, Vt., private communication.
[18] R. R. Troutman, "Subthreshold design considerations for insulated gate field-effect transistors," *IEEE J. Solid-State Circuits*, vol. SC-9, p. 55, April 1974.